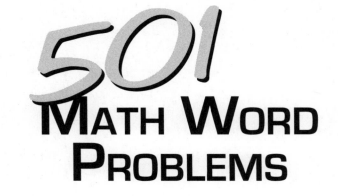

501 MATH WORD PROBLEMS

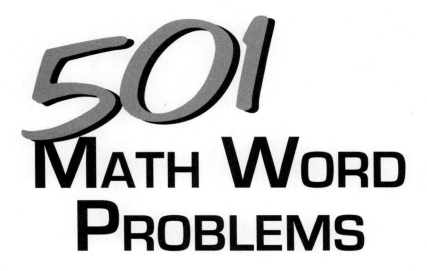

501 MATH WORD PROBLEMS

3rd Edition

Mark A. McKibben

NEW YORK

Printed in the United States of America

9 8 7 6 5 4 3 2 1

Third Edition

ISBN 978-1-57685-904-9

For more information or to place an order, contact LearningExpress at:
  2 Rector Street
  26th Floor
  New York, NY 10006

Or visit us at:
  www.learningpressllc.com

Contents

Introduction

Welcome to *501 Math Word Problems!* This book is designed to provide you with review and practice for math success. It provides 501 problems so you can flex your muscles with a variety of mathematical concepts. *501 Math Word Problems* is designed for many audiences. It is for anyone who has ever taken a math course and needs to refresh and revive forgotten skills. It can be used to supplement current instruction in a math class. Or, it can be used by teachers and tutors who need to reinforce student skills. If at some point you feel you need further explanation about some of the more advanced math topics highlighted in this book, you can find them in other LearningExpress publications. *Algebra Success in 20 Minutes a Day, 501 Algebra Questions, Geometry Success in 20 Minutes a Day*, and *501 Geometry Questions* can provide more detail on these topics.

How to Use This Book

First, look at the table of contents to see the math topics covered in this book. The book is divided into six sections: Miscellaneous Math, Fractions, Decimals, Percents, Algebra, and Geometry. The structure follows a common sequence of math concepts. You may want to follow the sequence because the concepts become more advanced as the book progresses. However, if your skills are just rusty, or if you are using this book to supplement topics you are currently learning, you may want to jump around from topic to topic.

As you complete the math problems in this book, you will undoubtedly want to check your answers against the answer explanation section at the end of each chapter. Every problem in *501 Math Word Problems* has a complete answer explanation. For problems that require more than one step, a thorough step-by-step explanation is provided. This will help you understand the problem-solving process. The purpose of drill and skill practice is to make you proficient at solving problems. Like an athlete preparing for the next season or a musician warming up for a concert, you become skilled with practice. If, after completing all the problems in a section, you feel you need more practice, do the problems again. It's not the answer that matters most—it's the process and the reasoning skills that you want to master.

You will probably want to have a calculator handy as you work through some of the sections. It's always a good idea to use it to check your calculations. If you have difficulty factoring numbers, the multiplication chart on the next page may help you. If you are unfamiliar with prime numbers, use the list on the next page so you won't waste time trying to factor numbers that can't be factored. And don't forget to keep lots of scrap paper on hand.

Make a Commitment

Success does not come without effort. Make the commitment to improve your math skills, and work for understanding. *Why* you perform a math operation is just as important as *how* you do it. If you truly want to be successful, make a commitment to spend the time you need to do a good job. You can do it! When you achieve math success, you have laid the foundation for future challenges and success. So sharpen your pencil and practice!

Multiplication Table

×	2	3	4	5	6	7	8	9	10	11	12
2	4	6	8	10	12	14	16	18	20	22	24
3	6	9	12	15	18	21	24	27	30	33	36
4	8	12	16	20	24	28	32	36	40	44	48
5	10	15	20	25	30	35	40	45	50	55	60
6	12	18	24	30	36	42	48	54	60	66	72
7	14	21	28	35	42	49	56	63	70	77	84
8	16	24	32	40	48	56	64	72	80	88	96
9	18	27	36	45	54	63	72	81	90	99	108
10	20	30	40	50	60	70	80	90	100	110	120
11	22	33	44	55	66	77	88	99	110	121	132
12	24	36	48	60	72	84	96	108	120	132	144

Prime Numbers < 1,015

2	3	5	7	11	13	17	19	23	29
31	37	41	43	47	53	59	61	67	71
73	79	83	89	97	101	103	107	109	113
127	131	137	139	149	151	157	163	167	173
179	181	191	193	197	199	211	223	227	229
233	239	241	251	257	263	269	271	277	281
283	293	307	311	313	317	331	337	347	349
353	359	367	373	379	383	389	397	401	409
419	421	431	433	439	443	449	457	461	463
467	479	487	491	499	503	509	521	523	541
547	557	563	569	571	577	587	593	599	601
607	613	617	619	631	641	643	647	653	659
661	673	677	683	691	701	709	719	727	733
739	743	751	757	761	769	773	787	797	809
811	821	823	827	829	839	853	857	859	863
877	881	883	887	907	911	919	929	937	941
947	953	967	971	977	983	991	997	1,009	1,013

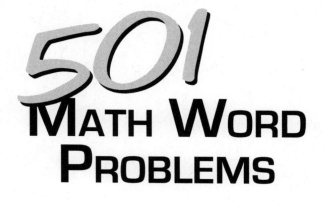

Miscellaneous Math

This chapter consists of 63 problems dealing with basic math concepts, including whole numbers, negative numbers, exponents, and square roots. It will provide a warm-up session before you move on to more difficult problems.

1. Bonnie has twice as many cousins as Robert. George has 5 cousins, which is 9 fewer than Bonnie has. How many cousins does Robert have?
   a. 17
   b. 22
   c. 4
   d. 7

2. Oscar sold 2 plasma televisions for every 5 DLP televisions he sold. If he sold 10 plasma televisions, how many DLP televisions did he sell?
   a. 45
   b. 20
   c. 25
   d. 10

3. Justin earned scores of 85, 92, and 95 on his science tests. What does he need
   to earn on his next science test so that the average (arithmetic mean) of the
   four exam scores is 93?
   a. 93
   b. 100
   c. 85
   d. 96

4. Brad's class collected 270 used books for the book drive. They packed
   them in boxes of 30 books each. How many boxes did they need?
   a. 240
   b. 10
   c. 9
   d. 5

5. Joey participated in a dance-a-thon. His team started dancing at 10 A.M.
   on Friday and stopped at 6 P.M. on Saturday. How many hours did Joey's
   team dance?
   a. 52
   b. 56
   c. 30
   d. 32

6. If one American dollar is worth 105 Japanese yen, how many dollars is 945
   yen worth?
   a. 6 dollars
   b. 9 dollars
   c. 1,000 dollars
   d. 99,225 dollars

7. Callie's grandmother pledged $0.85 for every mile Callie walked in her
   walk-a-thon. Callie walked 11 miles. How much money does her grandmother owe?
   a. $9.35
   b. $19.00
   c. $10.20
   d. $8.50

8. Tania needs 2 cups of pumpkin to make a pumpkin spice cake. The can of pumpkin that she has contains 8 cups of pumpkin. How many cakes can she make?

   a. 4 cakes

   b. $\frac{1}{4}$ of a cake

   c. 16 cakes

   d. 8 cakes

9. Mr. Brown plowed 6 acres in 1 hour. At this rate, how long will it take him to plow 21 acres?

   a. 3 hours

   b. 4 hours

   c. 3.5 hours

   d. 4.75 hours

10. The price of a box of one dozen donuts doubled over a period of 4 years. Assuming the price continues to double every 4 years and that the price of a box of one dozen donuts was $1.50 in 2003, how much does a box of one dozen donuts cost in 2011?

   a. $4.50

   b. $6.00

   c. $8.00

   d. $12.00

11. A piece of cardboard is 2 mm thick. Suppose it was folded in half, then folded in half again, and folded in half once more. How thick is the folded piece of cardboard?

   a. 4 mm

   b. 8 mm

   c. 16 mm

   d. 32 mm

12. The low temperature in Anchorage, Alaska today was ⁻8 °F. The low temperature in Los Angeles, California was 63 °F. What is the difference between the two low temperatures?

   a. 55 °F

   b. 71 °F

   c. 61 °F

   d. 14 °F

13. The Robin's Nest Nursing Home had a fundraising goal of $9,500. By the end of the fundraiser, they had exceeded their goal by $2,100. How much did the nursing home raise?

    a. $7,400

    b. $13,600

    c. $10,600

    d. $11,600

14. Mount Everest is 29,028 feet high. Mount Kilimanjaro is 19,340 feet high. How much taller is Mount Everest?

    a. 9,688 feet

    b. 9,788 feet

    c. 11,347 feet

    d. 6,288 feet

15. The area of a square is 36 cm². What is the length of one side of the square?

    a. 6 cm

    b. 9 cm

    c. 18 cm

    d. 24 cm

16. Mark's car gets 30 miles per gallon, and Jodi's car gets 21 miles per gallon. When traveling from their home to a theater in Pittsburgh, PA, they both used a whole number of gallons of gas. What is the shortest distance that their house could be from the theater in Pittsburgh?

    a. 420 miles

    b. 300 miles

    c. 630 miles

    d. 210 miles

17. Lucy's youth group raised $1,569 for charity. The group decided to split the money evenly among 3 charities. How much money will each charity receive?

    a. $784.50

    b. $423.00

    c. $523.00

    d. $341.00

18. Jason made 9 two-point baskets and 3 three-point baskets in Friday's basketball game. He did not score any other points. How many points did he score?
   a. 21
   b. 12
   c. 24
   d. 27

19. Jeff left Hartford at 2:15 P.M. and arrived in Boston at 4:45 P.M. How long did the drive take him?
   a. 2.5 hours
   b. 2.3 hours
   c. 3.25 hours
   d. 2.75 hours

20. Shane rolls a die with faces numbered 1 through 6. What is the probability that Shane rolls a 3?
   a. $\frac{3}{6}$
   b. $\frac{1}{6}$
   c. $\frac{1}{3}$
   d. $\frac{1}{2}$

21. Susan traveled 114 miles in 2 hours. If she keeps going at the same rate, how long will it take her to complete the remaining 285 miles of her trip?
   a. 5 hours
   b. 3 hours
   c. 7 hours
   d. 4 hours

22. A flight from Baltimore to San Jose took 5 hours and covered 3,010 miles. What was the plane's average speed?
   a. 545 mph
   b. 612 mph
   c. 515 mph
   d. 602 mph

23. Larry purchased 3 pairs of pants for $18 each and 5 shirts for $24 each. How much did Larry spend?
   a. $42
   b. $54
   c. $174
   d. $186

24. How many square centimeters are in one square meter?
   a. 100 sq cm
   b. 10,000 sq cm
   c. 144 sq cm
   d. 100,000 sq cm

25. Raul's kitchen is 5 yards long. How many inches long is the kitchen?
   a. 180 inches
   b. 60 inches
   c. 500 inches
   d. 5,000 inches

26. Jeff burns 500 calories per hour bicycling. How long will he have to ride to burn 1,125 calories?
   a. 3 hours
   b. 2.5 hours
   c. 2.25 hours
   d. 2 hours

27. The temperature at 6 P.M. was 31°F. By midnight, it had dropped 40 °F. What was the temperature at midnight?
   a. 9 °F
   b. ⁻9 °F
   c. ⁻11°F
   d. 0 °F

28. The total ticket sales for a soccer game were $1,260; 210 tickets were purchased. If all the tickets are the same price, what was the cost of a ticket?
   a. $6.00
   b. $3.50
   c. $10.00
   d. $7.50

29. Sherman took his pulse for 10 seconds and counted 11 beats. What is Sherman's pulse rate in beats per minute?
  a. 210 beats per minute
  b. 110 beats per minute
  c. 66 beats per minute
  d. 84 beats per minute

30. Jennifer flipped a coin four times and got heads each time. What is the probability that she gets heads on the next flip?
  a. 1
  b. $\frac{1}{32}$
  c. $\frac{1}{2}$
  d. 0

31. Jody's English quiz scores are 56, 93, 72, 89, and 87. What is the median of her scores?
  a. 72
  b. 87
  c. 56
  d. 85.6

32. A gambler in Las Vegas lost $550, then won $875, then won another $430, and then lost $1,000. What was the gambler's overall gain or loss?
  a. ⁻$245
  b. $245
  c. ⁻$2,855
  d. $2,855

33. Twelve coworkers go out for lunch together and order three pizzas. Each pizza is cut into eight slices. If each person gets the same number of slices, how many slices will each person get?
  a. 4
  b. 3
  c. 5
  d. 2

34. Marvin is helping his teachers plan a field trip. There are 136 people going on the field trip, and each school bus holds 51 people. What is the minimum number of school buses they will need to reserve for the trip?
  a. 3
  b. 2
  c. 4
  d. 5

35. The stock market is extremely volatile, or goes up and down frequently. A certain stock price gained 12 points, then dropped 8 points, then gained 1 point, then dropped 6 points, and finally gained 5 additional points right before the closing bell. What is the total gain or loss for this stock's price?
  a. 32 points
  b. ⁻32 points
  c. 4 points
  d. ⁻4 points

36. Lance has 70 cents, Margaret has three-fourths of a dollar, Guy has two quarters and a dime, and Bill has six dimes. Who has the most money?
  a. Lance
  b. Margaret
  c. Guy
  d. Bill

37. The students at Norton School were asked to name their favorite type of pet. Of the 430 students surveyed, 258 said that their favorite type of pet was a dog. Suppose that only 100 students were surveyed, with similar results. About how many students in this survey would say that a dog is their favorite type of pet?
  a. 58
  b. 60
  c. 72
  d. 46

38. A group of five friends went out to lunch. The total bill for the lunch was $53.75. Their meals all cost about the same, so they wanted to split the bill evenly. Without considering tip, how much should each friend pay?
 a. $11.25
 b. $12.85
 c. $10.75
 d. $11.50

39. The value of a computer depreciates evenly over five years for tax purposes (meaning that every year the computer is worth less by the same amount and that it is worth $0 after five years). If a business paid $2,400 for a computer, how much will it have depreciated after 2 years?
 a. $480
 b. $1,200
 c. $820
 d. $960

40. A baker cools bagels on trays that can fit 11 rows of 5 bagels each. He has a cart on which he can store 9 such trays. What is the largest number of bagels that can be cooled at once if 3 such carts are filled to capacity?
 a. 55
 b. 495
 c. 1,485
 d. 2,970

41. A national park keeps track of how many people are in each car that enters the park. Today, 57 cars had 4 people, 61 cars had 2 people, 9 cars had 1 person, and 5 cars had 5 people. What is the average number of people per car? Round to the nearest person.
 a. 2
 b. 3
 c. 4
 d. 5

42. A large pipe dispenses 750 gallons of water in 50 seconds. At this rate, how long will it take to dispense 360 gallons?
    a. 14 seconds
    b. 33 seconds
    c. 24 seconds
    d. 27 seconds

43. The light on a lighthouse blinks 45 times a minute. How long will it take the light to blink 405 times?
    a. 11 minutes
    b. 4 minutes
    c. 9 minutes
    d. 6 minutes

44. A six-sided die is rolled and a coin is tossed. What is the probability that a 3 will be rolled and a tail tossed?
    a. $\frac{1}{2}$
    b. $\frac{1}{6}$
    c. $\frac{1}{12}$
    d. $\frac{1}{8}$

45. Wendy has 5 pairs of pants and 7 shirts. How many different outfits can she make with these items (each outfit consists of one pair of pants and one shirt)?
    a. 12
    b. 24
    c. 35
    d. 21

46. Audrey measured the width of her dining room in inches. It is 150 inches. How many feet wide is her dining room?
    a. 12 feet
    b. 9 feet
    c. 12.5 feet
    d. 10.5 feet

47. Sharon wants to make 15 half-cup servings of soup. How many ounces of soup does she need?
   a. 60 ounces
   b. 150 ounces
   c. 120 ounces
   d. 7.5 ounces

48. Justin weighed 8 lb 12 oz when he was born. At his two-week check-up, he had gained 8 ounces. What was his weight at the two-week check-up in pounds and ounces?
   a. 9 lb
   b. 8 lb 15 oz
   c. 9 lb 4 oz
   d. 10 lb 2 oz

49. One inch equals 2.54 centimeters. The dimensions of a table made in Europe are 85 cm wide by 120 cm long. What is the width of the table in inches? Round to the nearest tenth of an inch.
   a. 30 inches
   b. 215.9 inches
   c. 33.5 inches
   d. 47.2 inches

50. A bag contains 3 red, 6 blue, 5 purple, and 2 orange marbles. One marble is selected at random. What is the probability that the marble chosen is blue?
   a. $\frac{4}{13}$
   b. $\frac{3}{8}$
   c. $\frac{3}{16}$
   d. $\frac{3}{5}$

51. The operator of an amusement park game kept track of how many tries it took participants to win the game. The following is the data from the first ten people:

    2, 6, 3, 4, 6, 2, 8, 4, 3, 5

    What is the median number of tries it took these participants to win the game?
    a. 8
    b. 6
    c. 4
    d. 2

52. Max goes to the gym every fourth day. Ellen's exercise routine is to go every third day. Today is Monday, and both Max and Ellen are at the gym. What will the day of the week be the next time they are both at the gym?
    a. Sunday
    b. Wednesday
    c. Friday
    d. Saturday

53. Danny is a contestant on a TV game show. If he gets a question right, the points for that question are added to his score. If he gets a question wrong, the points for that question are subtracted from his score. Danny currently has 200 points. If he gets a 300-point question wrong, what will his score be?
    a. ⁻100
    b. 0
    c. ⁻200
    d. 100

54. Nicole jogs for 60 minutes four times a week, lifts weights for 45 minutes three times a week, attends a 90-minute Zumba class twice a week, and plays racquetball for 75 minutes twice a week. How long does Nicole spend exercising each week?
    a. 4 hours 30 minutes
    b. 11 hours 45 minutes
    c. 9 hours
    d. 6 hours 15 minutes

55. The Ravens played 25 home games this year. They had 11 losses and 2 ties. How many games did they win?
   a. 12
   b. 13
   c. 14
   d. 11

56. Nancy has 24 CDs, and each contains between 9 and 15 songs, inclusive. Which of the following is the correct range for the total number of songs in her collection?
   a. Between 216 and 360, inclusive
   b. Between 186 and 300, inclusive
   c. Between 33 and 49, inclusive
   d. Between 240 and 336, inclusive

57. Determine the next number in the following pattern.
   320, 160, 80, 40, . . .
   a. 35
   b. 30
   c. 10
   d. 20

58. Greg bought three boxes of holiday cards that each cost $9.00, two packages of red pens that each cost $3.00, and a package of green pens for $2.00. Assuming tax has been included in the prices, if he gives the clerk a $50 bill, how much change will he receive?
   a. $18.00
   b. $31.00
   c. $36.00
   d. $15.00

59. Which expression below is equal to 5?
   a. $(1 + 2)^2$
   b. $9 - 2^2$
   c. $11 - 10 \times 5$
   d. $45 \div 3 \times 3$

60. A bus picks up a group of tourists at a hotel. The sightseeing bus travels 2 blocks north, 2 blocks east, 1 block south, 2 blocks east, and 1 block south. Where is the bus in relation to the hotel?
   a. 2 blocks north
   b. 1 block west
   c. 3 blocks south
   d. 4 blocks east

61. Each week Jaime saves $25. How long will it take her to save $350?
   a. 12 weeks
   b. 14 weeks
   c. 16 weeks
   d. 18 weeks

62. Ashley's car insurance costs her $115 per month. How much does it cost her per year?
   a. $1,150
   b. $1,380
   c. $980
   d. $1,055

63. The ratio of boys to girls at the dance was 3:4. There were 60 girls at the dance. How many boys were at the dance?
   a. 45
   b. 50
   c. 55
   d. 40

Answer Explanations

1. **d.** Work backwards to find the solution. George has 5 cousins, which is 9 fewer than Bonnie has; therefore, Bonnie has 14 cousins. Bonnie has twice as many cousins as Robert has, so half of 14 is 7. Robert has 7 cousins.

2. **c.** Set up a proportion with $\frac{plasma}{DLP}$: $\frac{2}{5} = \frac{10}{x}$. Cross-multiply and solve: (5)(10) = 2x. Divide both sides by 2: $\frac{50}{2} = \frac{2x}{2}$; $x = 25$ DLP televisions.

3. **b.** To earn an average of 93 on four tests, the sum of those four scores must be (93)(4), or 372. The sum of the first three scores is 85 + 92 + 95 = 272. The difference between the needed sum of the four scores and the sum of the first three scores is 100. Justin needs a 100 to earn a 93 average overall.

4. **c.** To find the number of boxes needed, you should divide the number of books by 30 books per box; 270 ÷ 30 = 9 boxes.

5. **d.** From 10 A.M. Friday to 10 A.M. Saturday is 24 hours. Then, from 10 A.M. Saturday to 6 P.M. Saturday is another 8 hours. Together, that makes 32 hours.

6. **b.** Set up a proportion with $\frac{dollar}{yen}$: $\frac{1}{105} = \frac{x}{945}$. Cross-multiply and solve: 945 = 105x. Divide both sides by 105: $\frac{945}{105} = \frac{105x}{105}$; $x = 9$ dollars.

7. **a.** Multiply the number of miles (11) by the amount pledged per mile ($0.85): 11 × $0.85 = $9.35. To multiply decimals, multiply normally, then count the number of decimal places in the problem and move the decimal point of the answer that many places to the left.

8. **a.** Set up a proportion with $\frac{pumpkin}{cakes}$: $\frac{2\ cups}{1\ cake} = \frac{8\ cups}{x\ cakes}$. Cross-multiply and solve: 8 = 2x. Divide both sides by 2: $\frac{8}{2} = \frac{2x}{2}$; $x = 4$ cakes.

9. **c.** Mr. Brown can plow 6 acres in an hour, so divide the number of acres to plow (21) by 6 to find the number of hours needed: 21 ÷ 6 = 3.5 hours.

10. b. First, determine the number of years between 2003 and 2011 by sub-tracting: 2011 − 2003 = 8. There are two 4-year periods in 8 years. The price after the first 4-year period, in 2007, is 2 × $1.50 = $3.00. Then, it doubles again in 2011 to reach a price of 2 × $3.00 = $6.00.

11. c. Each time the cardboard is folded in half, the thickness doubles. The cardboard is folded in half three times. The first time it is folded in half, the thickness becomes 2 × (2 mm) = 4 mm. The second time it is folded in half, the thickness becomes 2 × (4 mm) = 8 mm. Finally, folding it in half the third and final time yields a thickness of 2 × (8 mm) = 16 mm.

12. b. Visualize a number line. The distance from ‾8 to 0 is 8. Then, the distance from 0 to 63 is 63. Add the two distances together to get 71: 63 + 8 = 71. Alternatively, compute the difference 63 − (‾8) = 73.

13. d. *Exceeded* means "gone above." Therefore, if they exceeded their goal of $9,500 by $2,100, then they went over their goal by $2,100: $9,500 + $2,100 = $11,600. If you chose **a**, you subtracted $2,100 from $9,500 instead of adding the two numbers.

14. a. Subtract Mt. Kilimanjaro's height from Mt. Everest's height: 29,028 − 19,340 = 9,688. If you chose **b**, you did not borrow correctly when subtracting.

15. a. To find the area of a square, you multiply the length of one side by itself because all the sides are the same length. What number multiplied by itself is 36? 6 × 6 = 36.

16. d. Since we are told that they both used a whole number of gallons of gas, the smallest distance between their house and the theater is the *least common multiple* (LCM) of 21 and 30; this is the smallest whole number into which both 21 and 30 divide evenly. The LCM of 21 and 30 is 210. So, the shortest possible distance between their house and the theater is 210 miles.

17. c. Divide the money raised by three to find the amount that each charity will receive: $1,569 ÷ 3 = $523.00.

18. **d.** Find the number of points scored on two-point baskets by multiplying 2 × 9; 18 points were scored on two-point baskets. Find the number of points scored on three-point baskets by multiplying 3 × 3; 9 points were scored on three-point baskets. The total number of points scored is the sum of these two totals: 18 + 9 = 27.

19. **a.** From 2:15 P.M. to 4:15 P.M. is 2 hours. Then, from 4:15 P.M. to 4:45 P.M. is another half hour. This is a total of 2.5 hours.

20. **b.** There is a 1 in 6 chance of rolling a 3 because there are 6 possible outcomes on a die, but only 1 outcome is a 3.

21. **a.** Find the rate at which Susan is traveling by dividing her distance by time: 114 miles ÷ 2 hours = 57 mph. To find out how long it will take her to travel 285 miles, divide this distance by her rate: 285 miles ÷ 57 mph = 5 hours.

22. **d.** Divide the miles by the time to find the average speed: 3,010 ÷ 5 = 602 mph.

23. **c.** He spent $54 on pants (3 × $18 = $54) and $120 on shirts (5 × $24 = $120). Altogether he spent $174 ($54 + $120 = $174). If you chose **a**, you calculated the cost of *one* pair of pants plus *one* shirt instead of *three* pants and *five* shirts.

24. **b.** There are 100 cm in one meter. A square meter is 100 cm by 100 cm. The area of this is 10,000 sq cm (100 × 100 = 10,000).

25. **a.** There are 36 inches in a yard: 5 × 36 = 180 inches. There are 180 inches in 5 yards.

26. **c.** To find the number of hours needed to burn 1,250 calories, divide 1,250 by 500: 1,125 ÷ 500 = 2.25 hours.

27. **b.** Visualize a number line. The drop from 31° to 0° is 31°. There are still 9 more degrees to drop. This causes the temperature to drop below zero. ⁻9 °F is the temperature at midnight.

28. **a.** Divide the total sales ($1,260) by the number of tickets sold (210) to find the cost per ticket: $1,260 ÷ 210 = $6.00.

29. c. A 10 second count is $\frac{1}{6}$ of a minute. To find the number of beats per minute, multiply the beat in 10 seconds by 6: $11 \times 6 = 66$ beats per minute.

30. c. The probability of getting heads does not change based on the results of previous flips because each flip is an independent event. Therefore, the probability of getting heads is $\frac{1}{2}$.

31. b. To find the median, first put the numbers in order from least to greatest: 56, 72, 87, 89, 93. The middle number is the median. 87 is in the middle of the list, so it is the median. If you chose **a,** you forgot to put the numbers in ascending order before finding the middle number.

32. a. A negative integer describes a loss, and a positive integer describes a gain. Using this, the gambler's net gain or loss is computed using the following sum: $-\$550 + \$875 + \$430 + -\$1,000 = -\$245$.

33. d. Find the total number of slices by multiplying 3 by 8: $3 \times 8 = 24$. There are 24 slices to be shared among 12 coworkers. Divide the number of slices by the number of people to find the number of slices per person: $24 \div 12 = 2$ slices per person.

34. a. Divide the number of people by the number that fit on one bus: $136 \div 51 = 2.667$. They need more than 2 buses, but not quite 3. Since they can't order part of a bus, they will need to order 3 buses.

35. c. A negative integer describes a loss, and a positive integer describes a gain. Using this, the net gain or loss in the stock's price is computed using the following sum: $12 + -8 + 1 + -6 + 5 = 4$. So, the net gain in the stock's price is 4 points.

36. b. Lance has 70 cents. Margaret has three-fourths of a dollar, which is 75 cents. Guy has 60 cents ($25 + 25 + 10 = 60$). Bill has 60 cents ($6 \times 10 = 60$). So, Margaret has the most money.

37. b. Finding what 100 students would say is the same as finding the percent, because *percent* means "out of 100." To find the percent, divide the number of students who said a dog was their favorite (258) by the total number of students surveyed (430): 258 ÷ 430 = 0.6. Change 0.6 to a percent by moving the decimal two places to the right: 60%. This means that 60 out of 100 students would say that a dog is their favorite type of pet.

38. c. Divide the bill by 5: $53.75 ÷ 5 = $10.75. They should each pay $10.75.

39. d. Find how much the computer depreciates over one year by dividing the cost by 5: $2,400 ÷ 5 = $480. Multiply this by 2 for two years: $480 × 2 = $960. The computer will have depreciated $960 after 2 years.

40. c. First, a single tray can hold 5 × 11 = 55 bagels. Since a cart can hold 9 trays, the total number of bagels that can be cooled if a single cart is filled to capacity is 9 × 55 = 495 bagels. Since there are 3 carts that can be used, the largest number of bagels that can be cooled if all three carts are filled to capacity is 3 × 495 = 1,485 bagels.

41. b. Find the total number of people and the total number of cars. Then, divide the total people by the total cars.

People:	57 × 4 =	228
	61 × 2 =	122
	9 × 1 =	9
	5 × 5 =	25
	TOTAL	384 people

Cars: 57 + 61 + 9 + 5 = 132 cars

384 ÷ 132 = 2.9, which is rounded up to 3 people because 2.9 is closer to 3 than it is to 2.

42. c. Find the number of gallons dispensed per second by dividing 750 by 50 (750 ÷ 50 = 15 gallons per second). Divide 360 gallons by 15 to find how many seconds it will take (360 ÷ 15 = 24 seconds). It will take 24 seconds.

43. c. Divide 405 blinks by 45 blinks per minute to get 9 minutes.

44. c. Find the probability of each event separately, and then multiply the answers because they are independent outcomes. The probability of rolling a 3 is $\frac{1}{6}$, and the probability of tossing a tail is $\frac{1}{2}$. To find the probability of both of them happening, multiply $\frac{1}{6} \times \frac{1}{2} = \frac{1}{12}$. The probability is $\frac{1}{12}$. Alternatively, list all of the possible outcomes in the form of an ordered pair (die, coin):

$(1, H), (2, H), (3, H), (4, H), (5, H), (6, H)$

$(1, T), (2, T), (3, T), (4, T), (5, T), (6, T)$

45. c. Multiply the number of choices for each item to find the number of outfits: $5 \times 7 = 35$. There are 35 possible outfit combinations.

46. c. There are 12 inches in a foot. Divide 150 by 12 to find the number of feet: $150 \div 12 = 12.5$ feet.

47. a. One cup is 8 ounces, so half a cup is 4 ounces. Multiply 15 by 4 ounces to find the number of ounces needed: $15 \times 4 = 60$ ounces.

48. c. There are 16 ounces in a pound. If Justin gains 8 ounces, his weight will be 8 pounds and 20 ounces. 20 ounces is 1 pound and 4 ounces, so add this to the 8 pounds to get 9 lbs and 4 oz.

49. c. Divide the width (85 cm) by 2.54 to find the number of inches: $85 \div 2.54 = 33.46$ inches. The question says to round to the nearest tenth (one decimal place), which would be 33.5 inches.

50. b. The probability of choosing a blue marble is $\frac{blue}{total}$. The number of blue marbles is 6, and the total number of marbles is 16 ($3 + 6 + 5 + 2 = 16$). Therefore, the probability of choosing a blue marble is $\frac{6}{16} = \frac{3}{8}$.

51. c. First, arrange the numbers in order from least to greatest, and then find the middle of the set.

2, 2, 3, 3, 4, 4, 5, 6, 6, 8

The middle is the average (mean) of the 5th and 6th data items. The mean of 4 and 4 is 4.

52. d. A chart like the one below can be used to determine which days Max and Ellen go to the gym. The first day after Monday that they both go is Saturday.

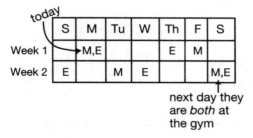

next day they are *both* at the gym

53. a. 200 – 300 = ⁻100 points

54. b. First, determine the total number of minutes Nicole spends exercising:

$$\underbrace{(60 \times 4)}_{\text{jogging}} + \underbrace{(45 \times 3)}_{\text{lifting weights}} + \underbrace{(90 \times 2)}_{\text{Zumba}} + \underbrace{(75 \times 2)}_{\text{raquetball}} = 705 \text{ minutes}$$

Now, convert this to hours and minutes using long division. Since there are 60 minutes in one hour, divide 705 by 60 to get 11 with remainder 45. So, she spends 11 hours 45 minutes exercising each week.

55. a. Thirteen games are accounted for with the losses and ties (11 ÷ 2 = 13). The remainder of the 25 games were won. Subtract to find the number of games won: 25 – 13 = 12 games won.

56. a. The smallest number of songs would occur if each of the 24 CDs had only 9 songs on it; this would yield 9 × 24 = 216 songs. The largest number of songs would occur if each of the 24 CDs had 15 songs on it; this would yield 15 × 24 = 360 songs. So, the total number of songs that Nancy has in her collection is between 216 and 360, inclusive.

57. d. Each number is divided by 2 to find the next number: 40 ÷ 2 = 20. Twenty is the next number.

58. d. Determine the total cost of all of the items Greg bought by multiplying the cost of each item by the number of that item that was purchased: $(3 \times \$9.00) + (2 \times \$3.00) + (1 \times \$2.00) = \$27.00 + \$6.00 + \$2.00 = \$35.00$ Now, subtract this total from $50.00 to determine the change Greg is due: $50.00 – $35.00 = $15.00.

59. b. The correct order of operations must be used here. PEMDAS tells you that you should do the operations in the following order: **P**arentheses, **E**xponents, **M**ultiplication and **D**ivision (left to right), **A**ddition and **S**ubtraction (left to right).

$9 - 2^2 = 9 - 4 = 5$

a is $(1 + 2)^2 = (3)^2 = 9$

c is $11 - 10 \times 5 = 11 - 50 = -39$

d is $45 \div 3 \times 3 = 15 \times 3 = 45$

60. d. See the diagram below. The bus is 4 blocks east of the hotel.

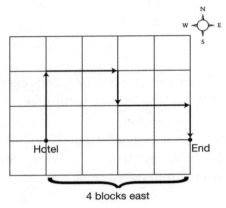

4 blocks east

61. b. Divide $350 by $25 per week: $350 \div 25 = 14$ weeks.

62. b. Multiply $115 by 12 because there are 12 months in a year: $\$115 \times 12 = \$1{,}380$ per year.

63. a. Use a proportion comparing boys to girls at the dance.

$$\frac{boys}{girls} = \frac{3}{4} = \frac{x}{60}$$

Solve the proportion by cross-multiplying, setting the cross-products equal to each other, and solving as shown below.

$(3)(60) = 4x$

$180 = 4x$

$\frac{180}{4} = \frac{4x}{4}$

$45 = x$

There were 45 boys at the dance.

Fractions

In order to understand arithmetic in general, it is important to practice and become comfortable with manipulating fractions. The problems in this chapter help you practice how to perform basic operations with fractions and will assist you in understanding real-world applications involving them.

64. Lori ran $5\frac{1}{2}$ miles Monday, $6\frac{1}{4}$ miles Tuesday, $4\frac{1}{2}$ miles Wednesday, and $2\frac{3}{4}$ miles on Thursday. What is the average number of miles Lori ran?

    a. 5

    b. $4\frac{1}{2}$

    c. 4

    d. $4\frac{3}{4}$

65. Last year Jonathan was $59\frac{5}{8}$ inches tall. This year he is $64\frac{3}{8}$ inches tall. How many inches did he grow?

    a. $5\frac{1}{2}$

    b. $4\frac{3}{4}$

    c. $4\frac{1}{4}$

    d. $5\frac{3}{4}$

66. Larry spends $\frac{3}{4}$ hour twice a day walking and playing with his dog. He also spends $\frac{1}{6}$ hour twice a day feeding his dog. How much time does Larry spend on activities involving his dog each day?
   a. $\frac{11}{12}$ hour
   b. $1\frac{1}{2}$ hours
   c. $1\frac{5}{6}$ hours
   d. $1\frac{2}{5}$ hours

67. The first section of a newspaper is comprised of 16 pages. Advertisements take up $3\frac{3}{8}$ of the pages. How many pages are not advertisements?
   a. $12\frac{5}{8}$
   b. $19\frac{3}{8}$
   c. 13
   d. $12\frac{1}{2}$

68. Lisa was assigned 48 pages to read for English class. She has finished $\frac{3}{4}$ of the assignment. How many more pages must she read?
   a. 36
   b. 21
   c. 12
   d. 8

69. Mark has three $4\frac{1}{2}$ oz cans of tomatoes and five $8\frac{1}{4}$ oz cans. How many ounces of tomatoes does Mark have?
   a. $12\frac{3}{4}$
   b. $54\frac{3}{4}$
   c. 54
   d. $62\frac{1}{4}$

70. Joe walked $2\frac{3}{4}$ miles to school, $1\frac{1}{3}$ miles to the library, and $1\frac{2}{5}$ miles to his friend's house. How many miles did Joe walk altogether?
   a. $4\frac{1}{2}$
   b. $5\frac{29}{60}$
   c. $5\frac{3}{4}$
   d. $4\frac{3}{5}$

71. Justin read $\frac{1}{8}$ of a book the first day, $\frac{1}{3}$ the second day, and $\frac{1}{4}$ the third day. On the fourth day he finished the book. What fraction of the book did Justin read on the fourth day?

    a. $\frac{2}{5}$

    b. $\frac{3}{8}$

    c. $\frac{7}{24}$

    d. $\frac{17}{24}$

72. Tammi babysat for $5\frac{1}{2}$ hours. She charged $7 an hour. How much should she get paid?

    a. $35.50

    b. $42

    c. $35

    d. $38.50

73. Tom needs to spread grass seed across his $3\frac{3}{4}$ acres of property. If he uses $2\frac{2}{3}$ bags of grass seed per acre, how many bags of grass seed does Tom need to purchase?

    a. 7

    b. 8

    c. 9

    d. 10

74. One batch of scones will provide enough for $\frac{3}{11}$ of Beth's Sunday brunch gathering of 44 people. If she bakes five batches of scones, how many additional people could she feed?

    a. 16

    b. 11

    c. 8

    d. 60

75. Allison baked a cake for Guy's birthday; $\frac{4}{7}$ of the cake was eaten at the birthday party. The next day, Guy ate one third of what was left. How much of the original cake did Guy eat the next day?

   a. $\frac{1}{7}$

   b. $\frac{3}{7}$

   c. $\frac{1}{4}$

   d. $\frac{4}{21}$

76. Josh practiced his clarinet for $\frac{5}{6}$ of an hour. How many minutes did he practice?

   a. 83

   b. 50

   c. 8.3

   d. 55

77. Tike Television has six minutes of advertising space every 15 minutes. How many $\frac{3}{4}$-minute commercials can be fit into the six-minute block?

   a. $4\frac{1}{2}$

   b. 8

   c. 20

   d. 11

78. Betty grew $\frac{3}{4}$ inch over the summer. Her friends also measured themselves and found that Susan grew $\frac{2}{5}$ inch, Mike grew $\frac{5}{8}$ inch, and John grew $\frac{1}{2}$ inch. List the friends in order of who grew the least to who grew the most.

   a. Betty, John, Mike , Susan

   b. Susan, Mike, John, Betty

   c. John, Mike, Betty, Susan

   d. Susan, John, Mike, Betty

79. An industrial-sized box of cereal contains 21 cups of cereal. If a serving size is $\frac{3}{4}$-cup, how many servings does the box contain?
    a. Nearly 16
    b. 28
    c. 21
    d. 14

80. The Grecos are taking an $8\frac{3}{4}$ mile walk. If they walk at an average speed of $3\frac{1}{2}$ miles per hour, how long will it take them?
    a. $2\frac{2}{3}$ hours
    b. $30\frac{1}{8}$ hours
    c. $2\frac{1}{2}$ hours
    d. 5 hours

81. The town's annual budget totals $32 million. If $\frac{2}{5}$ of the budget goes toward education, how many dollars go to education?
    a. $\$6\frac{1}{2}$ million
    b. $\$9\frac{1}{5}$ million
    c. $\$16$ million
    d. $\$12\frac{4}{5}$ million

82. A recipe that makes 4 dozen snickerdoodles requires $2\frac{1}{3}$ cups of flour. How many cups of flour would be needed to bake $2\frac{1}{2}$ times the recipe?
    a. $9\frac{1}{3}$
    b. $4\frac{2}{3}$
    c. $23\frac{1}{3}$
    d. $5\frac{5}{6}$

83. Linus wants to buy ribbon to make three bookmarks. One bookmark will be $11\frac{1}{2}$ inches long and the other two will be $8\frac{1}{4}$ inches long. How much ribbon should he buy?
    a. 27 inches
    b. 19 inches
    c. $19\frac{3}{4}$ inches
    d. 28 inches

84. Rita caught fish that weighed $3\frac{1}{4}$ lb, $8\frac{1}{2}$ lb, and $4\frac{2}{3}$ lb. What was the total weight of all the fish that Rita caught?
    a. $15\frac{4}{9}$ lb
    b. $14\frac{2}{3}$ lb
    c. $16\frac{5}{12}$ lb
    d. 15 lb

85. The dimensions of a rectangular garden are $4\frac{1}{2}$ yards by 3 yards. How many yards of fence are needed to surround the garden?
    a. $16\frac{1}{2}$
    b. $7\frac{1}{2}$
    c. 15
    d. 14

86. The soccer team is making pizzas for a fundraiser. They put $\frac{1}{4}$ of a package of cheese on each pizza. If they have 12 packages of cheese, how many pizzas can they make?
    a. 48
    b. 3
    c. 24
    d. $11\frac{3}{4}$

87. Dan purchased $3\frac{1}{2}$ yards of mulch for his garden. Mulch costs $25 a yard. How much did Dan pay for his mulch?
    a. $75.00
    b. $87.50
    c. $64.25
    d. $81.60

88. Ten tablespoons of instant iced tea mix are needed to make one pitcher of iced tea. One pitcher of iced tea can fill 6 glasses. How many tablespoons of instant iced tea mix are needed to fill $1\frac{1}{2}$ glasses?
    a. $1\frac{1}{9}$
    b. $2\frac{1}{2}$
    c. 3
    d. $3\frac{1}{4}$

89. Lucy worked $32\frac{1}{2}$ hours last week and earned \$195. What is her hourly wage?

   a. \$7.35

   b. \$5.00

   c. \$6.09

   d. \$6.00

90. A sheet of plywood is $5\frac{1}{2}$ feet wide and $7\frac{1}{2}$ feet long. What is the area of the sheet of plywood?

   a. $41\frac{1}{4}$ sq ft

   b. $82\frac{1}{2}$ sq ft

   c. $35\frac{1}{4}$ sq ft

   d. 11 sq ft

91. Rosa kept track of how many hours she spent reading during the month of August. The first week she read for $4\frac{1}{2}$ hours, the second week for $3\frac{3}{4}$ hours, the third week for $8\frac{1}{5}$ hours, and the fourth week for $1\frac{1}{3}$ hours. How many hours altogether did she spend reading in the month of August?

   a. $17\frac{47}{60}$

   b. 16

   c. $16\frac{1}{8}$

   d. $18\frac{2}{15}$

92. During a commercial break in the Super Bowl, there were four $\frac{1}{2}$-minute commercials and three $\frac{1}{4}$-minute commercials. How many minutes was the commercial break?

   a. $2\frac{3}{4}$

   b. $\frac{3}{4}$

   c. $3\frac{1}{2}$

   d. 5

93. Refinancing a home loan amounted to reducing the monthly payment by $\frac{1}{9}$ of the present monthly payment. If the new monthly payment is $1,128, how much was the original monthly payment?

a. $1,340
b. $1,213
c. $1,269
d. $1,225

94. Michelle is making a triple batch of chocolate chip cookies. The original recipe calls for $\frac{3}{4}$ cup of brown sugar. How many cups should she use if she is tripling the recipe?

a. $\frac{1}{4}$
b. 3
c. $2\frac{1}{4}$
d. $1\frac{3}{4}$

95. The output capacity of the equivalent of $1\frac{1}{3}$ printers is 15 pages per minute. How many pages would six printers, working together, produce in 8 minutes?

a. 120
b. 480
c. 540
d. 800

96. The Cheshire Senior Center is hosting a bingo night; $2,400 in prize money will be given away. The first winner of the night will receive $\frac{1}{3}$ of the money. The next ten winners will receive $\frac{1}{10}$ of the remaining amount. How many dollars will each of these ten winners receive?

a. $240
b. $800
c. $160
d. $200

97. $\frac{3}{8}$ of the employees in the Acme Insurance Company work in the accounting department. What fraction of the employees does NOT work in the accounting department?

    a. $\frac{10}{16}$

    b. $\frac{5}{16}$

    c. $\frac{3}{4}$

    d. $\frac{1}{2}$

98. A bag contains 44 cookies. Kyle eats 8 of the cookies. What fraction of the cookies is left?

    a. $\frac{5}{6}$

    b. $\frac{9}{11}$

    c. $\frac{8}{9}$

    d. $\frac{3}{4}$

99. Marci's car has a 12-gallon gas tank. Her fuel gauge says that there is $\frac{1}{8}$ of a tank left. How many gallons of gas are left in the tank?

    a. 3

    b. $1\frac{2}{3}$

    c. $\frac{3}{4}$

    d. $1\frac{1}{2}$

100. Joey, Aaron, Barbara, and Stu have been collecting pennies and putting them in identical containers. Joey's container is $\frac{3}{4}$ full, Aaron's is $\frac{3}{5}$ full, Barbara's is $\frac{2}{3}$ full, and Stu's is $\frac{2}{5}$ full. Whose container has the most pennies?

    a. Joey

    b. Aaron

    c. Barbara

    d. Stu

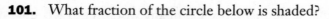

101. What fraction of the circle below is shaded?

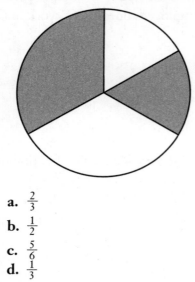

    a. $\frac{2}{3}$

    b. $\frac{1}{2}$

    c. $\frac{5}{6}$

    d. $\frac{1}{3}$

102. Michelle's brownie recipe calls for $2\frac{3}{4}$ cups of sugar. How much sugar does she need if she triples the recipe?

    a. $6\frac{1}{4}$ cups

    b. $7\frac{3}{4}$ cups

    c. $5\frac{1}{2}$ cups

    d. $8\frac{1}{4}$ cups

103. The output capacity of the equivalent of $1\frac{1}{4}$ printers is 20 pages per minute. The equivalent of how many printers would need to work together in order to produce 1,212 pages in three minutes?

    a. $25\frac{1}{4}$

    b. $15\frac{3}{4}$

    c. $12\frac{1}{6}$

    d. $22\frac{1}{2}$

104. Mr. Reynolds owns $1\frac{3}{4}$ acres of land. He plans to buy the property next to his, which is $2\frac{1}{8}$ acres. How many acres will Mr. Reynolds own after the purchase?

    a. $5\frac{1}{4}$

    b. $3\frac{3}{4}$

    c. $3\frac{1}{3}$

    d. $3\frac{7}{8}$

105. If Allen gets three oil changes in an eight-month period, how many oil changes will he have gotten at the end of $4\frac{2}{3}$ years?

    a. 7

    b. 14

    c. 21

    d. 28

106. How many eighths are in $4\frac{5}{8}$?

    a. 45

    b. 48

    c. 37

    d. 9

107. Kim is baking cookies for a large party and wants to double the recipe. The original recipe calls for $\frac{2}{3}$ cup of margarine. How many cups should she use for the double batch?

    a. $\frac{5}{6}$

    b. $1\frac{2}{3}$

    c. $1\frac{1}{3}$

    d. $\frac{4}{6}$

108. Mr. Johnson owns $4\frac{3}{4}$ acres. He sells one-third of his land. How many acres does he own after the sale?

    a. $2\frac{1}{4}$

    b. $2\frac{3}{16}$

    c. $3\frac{2}{3}$

    d. $3\frac{1}{6}$

109. Tim, Bob, and Fred return home for a short visit and, in anticipation, their mother bakes 12 dozen cookies. Tim arrived home first and ate $\frac{1}{9}$ of the entire batch of cookies. Bob arrived next and gathered up $\frac{3}{16}$ of what remained and took them to a party. Fred was the last to arrive home and took $\frac{4}{13}$ of what remained back to school with him. What fraction of cookies remained after Tim, Bob, and Fred had all taken their share?

   a. $\frac{7}{26}$

   b. $\frac{1}{2}$

   c. $\frac{2}{5}$

   d. $\frac{5}{16}$

110. The local firefighters are doing a "fill the boot" fundraiser. Their goal is to raise $3,500. After 3 hours, they have raised $2,275. Which statement below is accurate?

   a. They have raised 35% of their goal.

   b. They have $\frac{7}{20}$ of their goal left to raise.

   c. They have raised less than $\frac{1}{2}$ of their goal.

   d. They have raised more than $\frac{3}{4}$ of their goal.

111. Lori has half a pizza left over from dinner last night. For breakfast, she eats $\frac{1}{3}$ of the leftover pizza. What fraction of the original pizza remains after Lori eats breakfast?

   a. $\frac{1}{4}$

   b. $\frac{1}{6}$

   c. $\frac{1}{3}$

   d. $\frac{3}{8}$

112. A ball is dropped from a height of 108 feet. Each time it bounces, it rebounds to a height that is two-thirds of its previous height. What is the height of the ball after its fourth bounce?

   a. $1\frac{1}{3}$ feet

   b. 64 feet

   c. $10\frac{2}{3}$ feet

   d. $21\frac{1}{3}$ feet

113. A large coffee pot holds 120 cups. It is now about three-fifths full. About how many cups are currently in the pot?

    a. 24 cups

    b. 72 cups

    c. 48 cups

    d. 90 cups

114. An airport is backlogged with planes trying to land and has ordered planes to circle until they are told to land. A plane is using fuel at the rate of $9\frac{1}{2}$ gallons per hour, and it has $6\frac{1}{3}$ gallons left in its tank. How long can the plane continue to fly?

    a. $1\frac{1}{2}$ hours

    b. $\frac{2}{3}$ hours

    c. $2\frac{3}{4}$ hours

    d. $\frac{3}{4}$ hours

115. A carpenter receives measurements from a homeowner for a remodeling project. The homeowner lists the length of a room as $12\frac{3}{4}$ feet, but the carpenter would prefer to work in feet and inches. What is the measurement in feet and inches?

    a. 12 feet 9 inches

    b. 12 feet 8 inches

    c. 12 feet 6 inches

    d. 12 feet 3 inches

116. The state of Connecticut will pay two-thirds of the cost of a new school building. If the city of New Haven is building a school that will cost a total of $15,300,000, how much will the state pay?

    a. $7,750,000

    b. $5,100,000

    c. $10,200,000

    d. $1,020,000

117. One-fourth of an inch on a map represents 150 miles. The distance on the map from Springfield to Oakwood is $3\frac{1}{2}$ inches. How many miles is it from Springfield to Oakwood?

    a. 600 miles

    b. 2,100 miles

    c. 1,050 miles

    d. 5,250 miles

118. Robert brings a painting to the framing store to be framed. He chooses a frame with a 8 in by 10 in opening. The painting is $4\frac{1}{2}$ in by $6\frac{1}{2}$ in. A mat will be placed around the painting to fill the 8 in by 10 in opening. If the painting is perfectly centered, what will the width of the mat be on each side of the painting? See the diagram below.

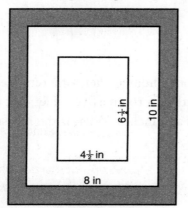

    a. $3\frac{1}{2}$ in

    b. $2\frac{1}{4}$ in

    c. $1\frac{1}{2}$ in

    d. $1\frac{3}{4}$ in

119. Mrs. Johnston's class broke into teams of three students each to play a game. The winning team received a $\frac{1}{2}$-pound jar of jellybeans. How many pounds of jellybeans will each team member get if the $\frac{1}{2}$-pound jar is shared equally among the three teammates?

    a. $\frac{1}{3}$ pound

    b. $\frac{1}{6}$ pound

    c. $\frac{2}{5}$ pound

    d. $\frac{2}{9}$ pound

120. Chuck is constructing a patio using $1\frac{1}{2}$ foot cement squares. The patio
will be 10 cement squares by 10 cement squares. If the cement squares are
placed right next to each other without any space in between, what will the
dimensions of the patio be?

a. 10 ft by 10 ft

b. 20 ft by 20 ft

c. $12\frac{1}{2}$ ft by $12\frac{1}{2}$ ft

d. 15 ft by 15 ft

121. Samantha's Girl Scout troop is selling holiday wreaths. Each wreath has
a bow that uses $\frac{2}{5}$ yard of ribbon. How many bows can Samantha make
from a spool of ribbon that is 10 yards long?

a. 25

b. 4

c. 20

d. 8

122. Lindsay and Mark purchased a $\frac{3}{4}$-acre plot of land to build a house. Zon-
ing laws require that houses built on less than 1 acre take up no more than
half the land. What is the largest amount of land that Lindsay and Mark's
house can cover?

a. $\frac{3}{8}$ acre

b. $\frac{1}{2}$ acre

c. $\frac{1}{4}$ acre

d. $\frac{5}{8}$ acre

123. Mr. Grove was watching the price of a stock he recently bought. On Monday, the stock was at $26\frac{3}{8}$. By Friday, the stock had fallen to $23\frac{1}{16}$. How
much did the stock price decline?

a. $2\frac{5}{8}$

b. $3\frac{5}{16}$

c. $1\frac{3}{16}$

d. $3\frac{1}{16}$

124. Four classmates are lined up in a row, and the first child in line is given a full 8 ounces of orange juice. He is instructed to pour $\frac{3}{4}$ of his glass into the glass of the next child in line. In turn, that child was instructed to pour $\frac{1}{2}$ of her glass into the glass of the next child in line. Finally, in turn, that child was instructed to pour $\frac{2}{3}$ of his glass's contents into the last child's glass. How many ounces of orange juice does the fourth child have in his glass?
   a. 1 ounce
   b. 2 ounces
   c. 3 ounces
   d. 4 ounces

125. Land in a develpment is setting for $60,000 per acre. If Jack purchases $1\frac{3}{4}$ acres, how much will be pay?
   a. $45,000
   b. $135,000
   c. $105,000
   d. $125,000

Answer Explanations

64. d. To find the average, add the miles Lori ran each day and divide by the number of days. To add the fractions, use a common denominator of 4: $5\frac{2}{4} + 6\frac{1}{4} + 4\frac{2}{4} + 2\frac{3}{4} = 17\frac{8}{4} = 17 + 2 = 19$. Divide the sum by 4: $19 \div 4 = 4\frac{3}{4}$.

65. b. Subtract to find the difference in heights. You will need to borrow 1 whole $\frac{8}{8}$ from 64 and add it to $\frac{3}{8}$ to make the fractional part of the mixed number $\frac{11}{8}$.
$64\frac{3}{8} - 59\frac{5}{8} = 63\frac{11}{8} - 59\frac{5}{8} = 4\frac{6}{8} = 4\frac{3}{4}$

If you chose **d**, you did not borrow, but rather simply subtracted the smaller fraction from the larger fraction.

66. c. Add the times together to find the total amount of time. Remember that he walks the dog twice and feeds the dog twice. The common denominator is 12.
$\frac{3}{4} + \frac{3}{4} + \frac{1}{6} + \frac{1}{6} = \frac{9}{12} + \frac{9}{12} + \frac{2}{12} + \frac{2}{12} = \frac{22}{12} = 1\frac{10}{12} = 1\frac{5}{6}$

If you chose **a**, you did not consider that he walks and feeds the dog *twice* a day.

67. a. Subtract the number of pages of advertisements from the total number of pages. Use a common denominator of 8 and borrow one whole ($\frac{8}{8}$) from 16 to do the subtraction.
$16 - 3\frac{3}{8} = 15\frac{8}{8} - 3\frac{3}{8} = 12\frac{5}{8}$

68. c. If Lisa has read $\frac{3}{4}$ of the assignment, she has $\frac{1}{4}$ left to go. To find $\frac{1}{4}$ of a number, divide the number by 4; $48 \div 4 = 12$ pages. If you chose **a**, you found the number of pages that she already read.

69. b. Ignore the fractional parts of the mixed numbers at first and multiply the whole number portion of the ounces by the corresponding number of cans: $4 \times 3 = 12$ ounces and $8 \times 5 = 40$ ounces. Adding together 12 and 40 gives a total of 52 ounces. Next, find the fractional portion. Multiply the fractional part by the corresponding number of cans: $\frac{1}{2} \times 3 = \frac{3}{2} = 1\frac{1}{2}$ ounces and $\frac{1}{4} \times 5 = \frac{5}{4} = 1\frac{1}{4}$ ounces. Add together these fractional parts: $1\frac{1}{2} + 1\frac{1}{4} = 2\frac{3}{4}$. Add this answer to the answer from the whole numbers to get the final answer: $2\frac{3}{4} + 52 = 54\frac{3}{4}$ ounces. If you chose **a**, you did not consider that Mark has *three* of the smaller cans and *five* of the larger cans.

70. b. To find the total distance Joe walked, add the three distances together using a common denominator of 60: $2\frac{45}{60} + 1\frac{20}{60} + 1\frac{24}{60} = 4\frac{89}{60}$, which is then simplified to $5\frac{29}{60}$.

71. c. First, find the fraction of the book that Justin read the first 3 days by adding the three fractions using a common denominator of 24: $\frac{3}{24} + \frac{8}{24} + \frac{6}{24} = \frac{17}{24}$. Subtract this fraction of the book from one whole, using a common denominator of 24: $\frac{24}{24} - \frac{17}{24} = \frac{7}{24}$. If you chose **d**, you found the fraction of the book that Justin read the first 3 days.

72. d. Multiply the hours Tammi babysat by the charge per hour. Change the mixed number to an improper fraction before multiplying: $\frac{11}{2} \times \frac{7}{1} = \frac{77}{2}$, which simplifies to $38\frac{1}{2}$ or $38.50.

73. d. Multiply the number of bags needed for one acre by the number of acres Tom has:
$3\frac{3}{4} \times 2\frac{2}{3} = \frac{15}{4} \times \frac{8}{3} = \frac{120}{12} = 10$ bags.

74. a. First, determine how many people corresponds to $\frac{3}{11}$ of the gathering by multiplying $\frac{3}{11}$ by 44: $\frac{3}{11} \times 44 = 12$ people. Next, let x stand for the number of people Beth can feed using 5 batches of scones. Set up the proportion $\frac{batches}{people}$ to solve for x: $\frac{1}{12} = \frac{5}{x}$. Cross multiply to solve for x: $60 = x$. So, five batches of scones will feed 60 people. Finally, to determine the number of people *beyond* the 44 attending the brunch gathering that Beth could feed, subtract 60 and 44: $60 - 44 = 16$ additional people.

75. a. Find the uneaten part of the cake by subtracting the eaten part from one whole: $1 - \frac{4}{7} = \frac{7}{7} = \frac{4}{7} = \frac{3}{7}$ of the cake was uneaten. To find one third of this amount, multiply by $\frac{1}{3}$: $\frac{3}{7} \times \frac{1}{3} = \frac{1}{7}$.

76. b. An hour is 60 minutes. To find the number of minutes in $\frac{5}{6}$ of an hour, multiply 60 by $\frac{5}{6}$: $\frac{60}{1} \times \frac{5}{6} = \frac{300}{6} = 50$ minutes.

77. b. Divide the 6-minute block by $\frac{3}{4}$, remembering to take the reciprocal of the second fraction and multiply: $6 \div \frac{3}{4} = \frac{6}{1} \times \frac{4}{3} = \frac{24}{3} = 8$.

78. d. To compare the fractions, use the common denominator of 40. Therefore, Betty = $\frac{30}{40}$, Susan = $\frac{16}{40}$, Mike = $\frac{25}{40}$, and John = $\frac{20}{40}$. To order the fractions, compare their numerators.

79. b. To find the number of servings, divide the number of cups of cereal contained within the box (21) by the fraction of a cup equivalent to a single serving ($\frac{3}{4}$): $21 \div \frac{3}{4} = 21 \times \frac{4}{3} = \frac{84}{3} = 28$ servings.

80. c. To find the amount of time that it took the Grecos to complete their walk, divide the distance ($8\frac{3}{4}$ miles) by the rate ($3\frac{1}{2}$ miles per hour). To divide mixed numbers, change them into improper fractions, then take the reciprocal of the second fraction and multiply:

$$8\frac{3}{4} \div 3\frac{1}{2} = \frac{35}{4} \div \frac{7}{2} = \frac{\cancel{35}^{5}}{\cancel{4}_{2}} \times \frac{\cancel{2}^{1}}{\cancel{7}_{1}} = \frac{5}{2} = 2\frac{1}{2} \text{ hours.}$$

81. d. To find $\frac{2}{5}$ of \$32 million, multiply the two numbers: $\frac{2}{5} \times \frac{32}{1} = \frac{64}{5}$: which simplifies to \$$12\frac{4}{5}$ million.

82. d. Multiply the number of cups of flour needed for a single recipe by the number of times you want to replicate the recipe: $2\frac{1}{3} \times 2\frac{1}{2} = \frac{7}{3} \times \frac{5}{2} = \frac{35}{6} = 5\frac{5}{6}$ cups of flour.

83. d. Add the needed lengths together using a common denominator of 4: $11\frac{2}{4} + 8\frac{1}{4} + 8\frac{1}{4} = 27\frac{4}{4}$, which simplifies to 28 inches.

84. c. Add the weights together using a common denominator of 12: $3\frac{3}{12} + 8\frac{6}{12} + 4\frac{8}{12} = 15\frac{17}{12}$, which simplifies to $16\frac{5}{12}$ lb.

85. c. Add all four sides of the garden together to find the perimeter. $4\frac{1}{2} + 4\frac{1}{2} + 3 + 3 = 14\frac{2}{2}$, which simplifies to 15 yards. If you chose **b**, you added only *two* sides of the garden.

86. a. Divide the number of packages of cheese (12) by $\frac{1}{4}$ to find the number of pizzas that can be made. Remember to take the reciprocal of the second number and multiply: $\frac{12}{1} \div \frac{1}{4} = \frac{12}{1} \times \frac{4}{1} = \frac{48}{1}$, which simplifies to 48. If you chose **b**, you multiplied by $\frac{1}{4}$ instead of dividing.

87. b. Multiply the number of yards purchased by the cost per yard. Change the mixed number into an improper fraction: $\frac{7}{2} \times \frac{25}{1} = \frac{175}{2}$, which reduces to $87\frac{1}{2}$ or \$87.50.

88. b. Let x stand for the number of tablespoons needed to make $1\frac{1}{2}$ glasses of iced tea. Set up a proportion of the form $\frac{teaspoons}{glasses}$: $\frac{10}{6} = \frac{x}{1\frac{1}{2}}$. Cross multiply and simplify: $10(1\frac{1}{2}) = 6x$; $10 \times \frac{3}{2} = 6x$; $15 = 6x$. Solve for x by dividing both sides by 6: $x = \frac{15}{6} = 2\frac{1}{2}$ tablespoons.

89. d. Divide the amount of money Lucy made by the number of hours she worked. Change the mixed number to an improper fraction. When dividing fractions, take the reciprocal of the second number and multiply: $195 \div 32\frac{1}{2} = \frac{195}{1} \div \frac{65}{2} = \frac{195}{1} \times \frac{2}{65} = \frac{390}{65}$, which simplifies to \$6.

90. a. To find the area, multiply the length by the width. When multiplying mixed numbers, change the mixed numbers to improper fractions: $5\frac{1}{2} \times 7\frac{1}{2} = \frac{11}{2} \times \frac{15}{2} = \frac{165}{4}$, which simplifies to $41\frac{1}{4}$ sq ft.

91. a. Add the number of hours together using a common denominator of 60: $4\frac{30}{60} + 3\frac{45}{60} + 8\frac{12}{60} + 1\frac{20}{60} = 16\frac{107}{60}$, which simplifies to $17\frac{47}{60}$ hours.

92. a. First, multiply 4 by $\frac{1}{2}$ to find the time taken by the four half-minute commercials: $4 \times \frac{1}{2} = 2$. Then, multiply $\frac{1}{4}$ by 3 to find the time taken by the three quarter-minute commercials: $\frac{1}{4} \times 3 = \frac{3}{4}$. Add the two times together to find the total commercial time. Use a common denominator of 4: $\frac{8}{4} + \frac{3}{4} = \frac{11}{4}$, which simplifies to $2\frac{3}{4}$ minutes.

93. c. Let x represent the amount of the original monthly payment. Since refinancing the home loan results in reducing the monthly payment by $\frac{1}{9}$, it follows that the *new* monthly payment $\frac{8}{9}$ of the *original* monthly payment, or $\frac{8}{9}x$. This is equal to \$1,128. Solve for x by dividing both sides by $\frac{8}{9}$: $x = \frac{\$1,128}{\frac{8}{9}} = \$1,128 \div \frac{8}{9} = \$1,128 \times \frac{9}{8} = \$1,269$.

94. c. Multiply the amount of brown sugar needed for one batch ($\frac{3}{4}$) by the number of batches (3): $\frac{3}{4} \times \frac{3}{1} = \frac{9}{4}$, which simplifies to $2\frac{1}{4}$ cups.

95. c. First, we must determine the number of pages *per minute* that six print-ers can produce when working together. Call this quantity x. To do so, set up a proportion of the form $\frac{printers}{pages}$: $\frac{1\frac{1}{3}}{15} = \frac{6}{x}$. Cross multiply and solve for x: $\left(1\frac{1}{3}\right)x = 15(6)$; $\frac{4}{3}x = 90$; $x = \frac{3}{4}(90) = 67\frac{1}{2}$ pages. So, in one minute, six printers can produce $67\frac{1}{2}$ pages. Finally, to deter-mine the number of pages produced in 8 minutes, multiply $67\frac{1}{2}$ by 8: $67\frac{1}{2} \times 8 = \frac{135}{2} \times 8 = 540$ pages.

96. c. The prize money is divided into tenths after the first third has been paid out. Find one third of $2,400 by dividing $2,400 by 3; $800 is paid to the first winner, leaving $1,600 for the next ten winners to split evenly ($2,400 – $800 = $1,600). Divide $1,600 by 10 to find how much each of the 10 winners will receive: $1,600 ÷ 10 = $160. Each of the last 10 win-ners will receive $160.

97. a. The entire company is one whole or $\frac{8}{8}$. Subtract $\frac{3}{8}$ from the whole to find the fraction of the company that does not work in accounting: $\frac{8}{8} - \frac{3}{8} = \frac{5}{8}$. Recall that you subtract the numerators and leave the denominators the same when subtracting then fractions; $\frac{5}{8}$ is equivalent to $\frac{10}{16}$.

98. b. If Kyle eats 8 cookies, then 36 cookies are left (44 – 8 = 36). The part that is left is 36, and the whole is 44. Therefore, the fraction is $\frac{36}{44}$. Both the numerator and denominator are divisible by 4. Divide both parts by 4 to simplify the fraction to $\frac{9}{11}$.

99. d. Find $\frac{1}{8}$ of 12 gallons by multiplying: $\frac{1}{8} \times \frac{12}{1} = \frac{12}{8}$. Recall that 12 is equivalent to 12 over 1. To multiply fractions, multiply the numerators and multiply the denominators. Change $\frac{12}{8}$ to a mixed number; 8 goes into 12 once, so the whole number is 1. Four remains, so $\frac{4}{8}$ or $\frac{1}{2}$ is the fractional part; $1\frac{1}{2}$ gallons are left.

100. a. Compare $\frac{3}{4}, \frac{3}{5}, \frac{2}{3}, \frac{2}{5}$ by finding a common denominator. The com-mon denominator for 3, 4, and 5 is 60. Multiply the numerator and denominator of a fraction by the same number so that the denominator becomes 60. The fractions then become $\frac{45}{60}, \frac{36}{60}, \frac{40}{60}, \frac{24}{60}$. The fraction with the largest numerator is the largest fraction; $\frac{45}{60}$ is the largest fraction. It is equivalent to Joey's fraction of $\frac{3}{4}$.

101. b. Break the circle into sixths as shown below; 3 of the sixths are shaded, which is equivalent to $\frac{3}{6}$ or $\frac{1}{2}$.

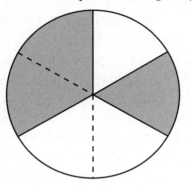

102. d. To triple the recipe, multiply by 3: $2\frac{3}{4} \times 3 = \frac{11}{4} \times \frac{3}{1} = \frac{33}{4} = 8\frac{1}{4}$. When multiplying a mixed number, change it to an improper fraction first. To find the numerator of the improper fraction, multiply the whole number by the denominator and add the product to the numerator. Keep the denominator the same.

103. a. First, let x represent the number of printers that could, working together, print 1,212 pages in 3 minutes. Such a collection of printers would print 404 pages per minute. Using this, set up a proportion of the form $\frac{printers}{pages\ per\ minutes}$: $\frac{1\frac{1}{4}}{20} = \frac{x}{404}$ Cross multiply and simplify: $\left(1\frac{1}{4}\right) \times 404 = 20x \frac{5}{4} \times 404 = 20x$; $505 = 20x$. Now, solve for x by dividing both sides by 20: $x = \frac{505}{20} = 25\frac{1}{4}$ printers.

104. d. The two pieces of land together are $1\frac{3}{4} + 2\frac{1}{8}$ acres. Add the whole numbers: $1 + 2 = 3$. Add the fractions separately writing each fraction with denominator 8: $\frac{3}{4} + \frac{1}{8} = \frac{6}{8} + \frac{1}{8} = \frac{7}{8}$. Add this to the whole number to get $3\frac{7}{8}$ acres.

105. c. First, since there are 12 months in one year, the number of months in $4\frac{2}{3}$ years is obtained by multiplying $4\frac{2}{3} \times 12$: $\frac{14}{3} \times 12 = 56$ months. Next, note that 56 months is comprised of seven 8-month periods, which is found by dividing 56 by 8. However, remember that Allen gets 3 oil changes in each 8-month period, so multiply 7 times 3 to conclude that there are 21 oil changes in this time frame.

106. c. Every whole has 8 eighths. Since there are 8 eighths in each of the 4 wholes, there are 32 eighths in the whole number portion. There are 5 eighths in the fraction portion. Adding the two parts together, you get 37 eighths (32 + 5 = 37).

107. c. To double the recipe, multiply the ingredients by 2: $\frac{2}{3} \times 2 = \frac{2}{3} \times \frac{2}{1} = \frac{4}{3}$. Recall that 2 can be written as $\frac{2}{1}$; $\frac{4}{3}$ is an improper fraction. To change it to a mixed number, determine how many times 3 goes into 4. It goes in one time, so the whole number is 1. There is one third left over, so the mixed number is $1\frac{1}{3}$.

108. d. If Mr. Johnson sells $\frac{1}{3}$ of his land, then he still owns $\frac{2}{3}$ of it. To find this, multiply $4\frac{3}{4}$ acres by $\frac{2}{3}$: $4\frac{3}{4} \times \frac{2}{3} = \frac{19}{4} \times \frac{2}{3} = \frac{19}{6} = 3\frac{1}{6}$ acres. In the second step of the multiplication, $4\frac{3}{4}$ was changed to an improper fraction, $\frac{19}{4}$. And in the last step, the improper fraction $\frac{19}{6}$ was converted to a mixed number.

109. b. There are 144 cookies in 12 dozen. First, determine the number of cookies Tim ate by computing $\frac{1}{9}$ of 144; multiply them together to get $\frac{1}{9} \times 144 = 16$. This leaves 144 − 16 = 128 cookies. Next, Bob takes $\frac{3}{16}$ of the remaining 128 cookies; this equals $\frac{3}{16} \times 128 = 24$ cookies. This leaves 128 − 24 = 104 cookies. Finally, Fred takes $\frac{4}{13}$ of the remaining 104 cookies; this equals $\frac{4}{13} \times 104 = 32$ cookies. This leaves 104 − 32 = 72 cookies. Thus, the fraction of the cookies that now remain is $\frac{72}{144} = \frac{1}{2}$.

110. b. The part of their goal that they have raised is $2,275, and the whole goal is $3,500. The fraction for this is $\frac{2,275}{3,500}$. The numerator and denominator can both be divided by 175 to get a simplified fraction of $\frac{13}{20}$. They have completed $\frac{13}{20}$ of their goal, which means that they have $\frac{7}{20}$ left to go ($\frac{20}{20} - \frac{13}{20} = \frac{7}{20}$).

111. c. Refer to the drawing below. If half is broken into thirds, each third is one-sixth of the whole. Therefore, she has $\frac{2}{6}$ or $\frac{1}{3}$ of the pizza left over.

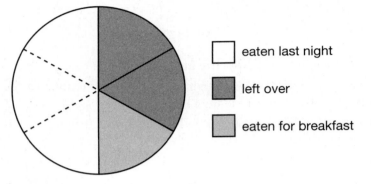

☐ eaten last night

▨ left over

▨ eaten for breakfast

112. d. The height of the ball after its first bounce is $\frac{2}{3} \times 108$ feet. The height of the ball after its second bounce is $\frac{2}{3}$ of the previous height, or $\frac{2}{3} \times (\frac{2}{3} \times 108)$ feet. Likewise, the height of the ball after the third bounce is $\frac{2}{3} \times (\frac{2}{3} \times (\frac{2}{3} \times 108))$ feet and after the fourth bounce is $\frac{2}{3} \times (\frac{2}{3} \times (\frac{2}{3} \times (\frac{2}{3} \times 108)))$ feet. Simplifying this last expression yields:

$$\frac{2}{3} \times \left(\frac{2}{3} \times \left(\frac{2}{3} \times \left(\frac{2}{3} \times 108\right)\right)\right) \text{ feet} = \frac{2 \times 2 \times 2 \times 2}{3 \times 3 \times 3 \times 3} \times 108 \text{ feet} = \frac{16}{81} \times 108 \text{ feet} = \frac{64}{3} \text{ feet} = 21\frac{1}{3} \text{ feet}.$$

113. b. Multiply 120 by $\frac{3}{5}$: $\frac{120}{1} \times \frac{3}{5} = \frac{360}{5} = 72$. 72 is written as a fraction with a denominator of 1. The fraction $\frac{360}{5}$ is simplified by dividing 360 by 5 to get 72 cups.

114. b. Use a proportion comparing gallons used to time. The plane uses $9\frac{1}{2}$ gallons in 1 hour and the problem asks how many hours it will take to use $6\frac{1}{3}$ gallons.

$$\frac{9\frac{1}{2}g}{1\,hr} = \frac{6\frac{1}{3}g}{x\,hrs}$$

To solve the proportion for x, cross multiply, set the cross-products equal to each other and solve as shown below.

$$(9\frac{1}{2})x = (6\frac{1}{3})(1)$$

$$\frac{9\frac{1}{2}x}{9\frac{1}{2}} = \frac{6\frac{1}{3}}{9\frac{1}{2}}$$

To divide the mixed numbers, change them into improper fractions.

$$x = \frac{\frac{19}{3}}{\frac{19}{2}}$$

$$x = \frac{19}{3} \div \frac{19}{2}$$

$$x = \frac{19}{3} \times \frac{2}{19}$$

The 19s cancel. Multiply the numerators straight across and do the same for the denominators. The final answer is $\frac{2}{3}$ hour.

115. a. There are 12 inches in a foot; $\frac{3}{4}$ of a foot is 9 inches ($\frac{3}{4} \times 12 = 9$). The length of the room is 12 feet 9 inches.

116. c. Multiply \$15,300,000 by $\frac{2}{3}$: $\frac{\$15,300,000}{1} \times \frac{2}{3} = \frac{\$30,600,000}{3} = \$10,200,000$.

117. b. Every $\frac{1}{4}$ inch on the map represents 150 miles in real life. There are 4 fourths in every whole ($4 \times 3 = 12$) and 2 fourths in $\frac{1}{2}$ for a total of 14 fourths ($12 + 2 = 14$) in $3\frac{1}{2}$ inches. Every fourth equals 150 miles. Therefore, there are 2,100 miles ($14 \times 150 = 2,100$) between Springfield and Oakwood.

118. d. The width of the opening of the frame is 8 inches. The painting only fills $4\frac{1}{2}$ inches of it. There is an extra $3\frac{1}{2}$ inches ($8 - 4\frac{1}{2} = 3\frac{1}{2}$) to be filled with the mat. There will be an even amount of space on either side of the painting. So, divide the extra space by 2 to find the amount of space on each side because it will be perfectly centered. To divide a mixed number by a whole number, change the mixed number to an improper fraction; $3\frac{1}{2}$ becomes $\frac{7}{2}$. Also, change 2 to a fraction ($\frac{2}{1}$). Then take the reciprocal of 2 and multiply: $\frac{7}{2} \times \frac{1}{2} = \frac{7}{4} = 1\frac{3}{4}$ inches.

119. b. Divide $\frac{1}{2}$ pound by 3. Recall that 3 can be written as $\frac{3}{1}$: $\frac{1}{2} \div \frac{3}{1}$. When dividing fractions, take the reciprocal of the second fraction and multiply: $\frac{1}{2} \times \frac{1}{3} = \frac{1}{6}$ pound of jellybeans per teammate.

120. d. Multiply $1\frac{1}{2}$ by 10. Change $1\frac{1}{2}$ to an improper fraction ($\frac{3}{2}$) and make 10 into a fraction by placing it over 1 ($\frac{10}{1}$): $\frac{3}{2} \times \frac{10}{1} = \frac{30}{2} = 15$ feet. Each side is 15 feet long, so the dimensions are 15 ft by 15 ft.

121. a. Divide 10 by $\frac{2}{5}$. Recall that 10 can be written as $\frac{10}{1}$: $\frac{10}{1} \div \frac{2}{5}$. When dividing fractions, take the reciprocal of the second fraction and multiply: $\frac{10}{1} \times \frac{5}{2} = \frac{50}{2} = 25$ bows. If you can't figure out what operation to use in this problem, consider what you would do if you had 10 yards of ribbon and each bow took 2 yards. You would divide 10 by 2. This is the same for this problem, except that the bow takes a fraction of a yard.

122. a. Multiply $\frac{3}{4}$ by $\frac{1}{2}$ to find half of the land: $\frac{3}{4} \times \frac{1}{2} = \frac{3}{8}$ acre.

123. b. Subtract Friday's price from Monday's price: $26\frac{3}{8} - 23\frac{1}{16}$. In order to subtract, you need a common denominator. The common denominator is 16. Multiply the first fraction by 2 in the numerator and 2 in the denominator: $\frac{3 \times 2}{8 \times 2} = \frac{6}{16}$. Then subtract: $26\frac{6}{16} - 23\frac{1}{16} = 3\frac{5}{16}$.

124. b. The idea each time is to multiply the fraction by the amount of orange juice in the glass of the child doing the pouring to get the amount in the next child's glass. To begin, the first child starts with 8 ounces of orange juice. The second child in line gets $\frac{3}{4}$ of this amount, which is $\frac{3}{4} \times 8$ ounces = 6 ounces. Next, the third child in line gets $\frac{1}{2}$ of the amount in the second child's glass, which is $\frac{1}{2} \times 6$ ounces = 3 ounces. Finally, the fourth child gets $\frac{2}{3}$ of the amount in the third child's glass, which is $\frac{2}{3} \times 3$ ounces = 2 ounces.

125. c. Multiply the cost per acre by the number of acres: $\$60,000 \times 1\frac{3}{4}$. Change 60,000 to a fraction by putting it over 1 and change the mixed number to an improper fraction: $\frac{\$60,000}{1} \times \frac{7}{4} = \frac{\$420,000}{4} = \$105,000$.

Decimals

We use decimals every day—to express amounts of money, or to measure distances and quantities. This chapter includes word problems that will help you practice your skills in rounding decimals, as well as performing basic mathematical operations with them.

126. A nearby star is approximately 2.3×10^{21} meters from Earth. One light year is equal to approximately 9.46×10^{15} meters. What is the approximate distance from Earth to this star measured in light years? Round to the nearest light year.
   a. 243,129 light years
   b. 243,128 light years
   c. 4,113,043 light years
   d. Less than 1 light year

127. Kara brought $26 with her when she went shopping. She spent $6.27 for lunch and $14.98 on a shirt. How much money does she have left?
   a. $8.02
   b. $4.75
   c. $18.25
   d. $7.38

128. Lucas purchased his motorcycle for $5,875.98 and sold it for $7,777.77. What was his profit?
a. $1,901.79
b. $2,902.89
c. $1,051.79
d. $1,911.89

129. Mike, Dan, Ed, and Sy played together on a baseball team. Mike's batting average was 0.349, Dan's was 0.2, Ed's was 0.35, and Sy's was 0.299. Who had the highest batting average?
a. Mike
b. Dan
c. Ed
d. Sy

130. Price Cutter sold 85 beach towels for $6.95 each. What were the total sales?
a. $865.84
b. $186.19
c. $90.35
d. $590.75

131. A computer depreciates to about 0.85 of its value from the year before. What was the original value, to the nearest cent, of a computer that is worth $650 at the end of four years?
a. $339.30
b. $399.18
c. $1,058.42
d. $1,245.20

132. Katie ran 11.1 miles over the last three days. How many miles did she average per day?
a. 3.7
b. 3.0
c. 2.4
d. 3.3

133. Sharon purchased six adult movie tickets. She spent $52.50 on the tickets. How much was each ticket?
   a. $5.25
   b. $8.25
   c. $7.00
   d. $8.75

134. A computer depreciates to about 0.85 of its value from the year before. If a computer is purchased for $2,200, how much will it be worth, to the nearest cent, in three years?
   a. $3,582.33
   b. $3,044.98
   c. $1,351.08
   d. $1,589.50

135. An international long distance call costs $0.50 per minute for each of the first ten minutes, and then $0.22 per minute for each minute thereafter. How much would a call lasting 1 hour 8 minutes cost?
   a. $14.96
   b. $16.44
   c. $17.76
   d. $34.00

136. Kenny used a micrometer to measure the thickness of a piece of construction paper. The paper measured halfway between 0.24 millimeters and 0.25 millimeters. What is the thickness of the paper?
   a. 0.05
   b. 0.245
   c. 0.255
   d. 0.3

137. In her last gymnastics competition Keri scored 5.65 on the floor exercise, 5.85 on the vault, and 5.75 on the balance beam. What was Keri's total score?
   a. 17.25
   b. 12.31
   c. 15.41
   d. 13.5

138. Linda bought 35 yards of fencing at $4.88 a yard. How much did she spend?
   a. $298.04
   b. $248.80
   c. $91.04
   d. $170.80

139. A group of friends rented an SUV and make a cross country trip with a total distance of 1,703.6 miles. In making the trip, $256.66 worth of gasoline at $3.13 per gallon was used. Approximately how many miles per gallon did they average in making this trip? Round to the nearest mile.
   a. 21 miles per gallon
   b. 24 miles per gallon
   c. 25 miles per gallon
   d. 29 miles per gallon

140. Using the fact that there are 2.54 cm in one inch, what are the dimensions of a 5 in. × 7 in. photo in cm? Express the dimensions to the hundredths place; round, if appropriate.
   a. 1.97 cm × 2.76 cm
   b. 2.0 cm × 2.8 cm
   c. 12.70 cm × 17.78 cm
   d. 12.7 cm × 17.8 cm

141. Last year's budget was $12.4 million. This year's budget is $14.3 million. How much did the budget increase?
   a. $2.5 million
   b. $1.9 million
   c. $1.4 million
   d. $2.1 million

142. Mrs. Hartill drove 3.1 miles to the grocery store, then 4.25 miles to the salon, and 10.8 miles to her son's house. How many miles did she drive altogether?
   a. 18.15
   b. 56.4
   c. 8.43
   d. 14.65

143. The following are four times from a 400-meter race. Which is the fastest time?
   a. 10.1
   b. 10.14
   c. 10.2
   d. 10.09

144. A man has just turned 38 years old, and has averaged 9 hours per 24-hour day sleeping. How many minutes of his life has he spent sleeping? Assume that each year consists of 365 days, ignoring the extra day that occurs in each leap year.
   a. 124,830
   b. 197,100
   c. 832,200
   d. 7,489,800

145. Joe's batting average is between 0.315 and 0.32. Which of the following could be Joe's average?
   a. 0.311
   b. 0.309
   c. 0.321
   d. 0.317

146. The professor-to-student ratio at a college of 1,450 students is approximately 1:16. How many additional professors must be hired to reduce the ratio to 1:12?
   a. 25
   b. 30
   c. 50
   d. 75

147. Brian's 100-yard dash time was 2.68 seconds more than the school record. Brian's time was 13.4 seconds. What is the school record?
   a. 10.72 seconds
   b. 11.28 seconds
   c. 10.78 seconds
   d. 16.08 seconds

148. Ryan's gym membership costs him $390 per year. He pays this in twelve equal installments a year. How much is each installment?
   a. $1,170
   b. $42.25
   c. $4,680
   d. $32.50

149. How much greater is 0.0101 than 0.003?
   a. 0.0401
   b. 0.071
   c. 0.0071
   d. 0.0131

150. If 9 ounces of turkey breast cost $4.80, how much does $2\frac{1}{2}$ pounds cost? Round your answer to the nearest cent.
   a. $12.00
   b. $16.00
   c. $17.07
   d. $21.33

151. Jay bought twenty-five $0.44 stamps. How much money did he spend?
   a. $8.44
   b. $9.74
   c. $8.80
   d. $11.00

152. During the summer months, gas prices increase weekly by about 0.06 of the previous week's cost. If one gallon of gas costs $2.80 at the beginning of the summer, what would the approximate cost per gallon be at the end of 2 weeks? Round your answer to the nearest cent.
   a. $2.92
   b. $2.97
   c. $3.14
   d. $3.15

153. Hanna's sales goal for the week is $5,000. So far she has sold $3,574.38 worth of merchandise. How much more money does she need to make to meet her goal?
   a. $2,425.38
   b. $1,329.40
   c. $2,574.38
   d. $1,425.62

154. The legend on a road map indicates that 1.5 inches equals 62.4 miles. Approximately how many inches would represent 473.8 miles? Round your answer to the nearest tenth.
   a. 7.6
   b. 5.1
   c. 11.4
   d. 12.5

155. Andy earned the following grades on his four math quizzes: 97, 78, 84, and 86. What is the average of his four quiz grades?
   a. 82.5
   b. 86.25
   c. 81.5
   d. 87

156. Luis runs at a rate of 11.7 feet per second. How far does he run in 5 seconds?
   a. 585 feet
   b. 490.65 feet
   c. 58.5 feet
   d. 55.5 feet

157. Suppose you invest in a company's stock at $23.81 per share. If you bought 250 shares, how much loss would you incur if the price per share decreased to $19.28?
   a. $1,132.50
   b. $2,410.00
   c. $4,820.00
   d. $5,952.50

158. Mike can jog 6.5 miles per hour. At this rate, how many miles will he jog in 1 hour and 30 minutes?
   a. 9.75 miles
   b. 4.7 miles
   c. 9 miles
   d. 10.75 miles

159. What decimal is represented by point A on the number line?

   a. 0.77
   b. 0.752
   c. 0.765
   d. 0.73

160. Nicole is making 20 gift baskets. She has 15 pounds of chocolates to distribute equally among the baskets. If each basket gets the same amount of chocolates, how many pounds should Nicole put in each basket?
   a. 1.3 pounds
   b. 0.8 pounds
   c. 0.75 pounds
   d. 3 pounds

161. A librarian is returning library books to the shelf. She uses the call numbers to determine where the books belong. She needs to reshelf a book about perennials with a call number of 635.93. Between which two call numbers should she place the book?
   a. 635.8 and 635.9
   b. 635.8 and 635.95
   c. 635.935 and 635.94
   d. 635.99 and 636.0

162. Michael made 19 out of 30 free-throws this basketball season. Larry's free-throw average was 0.745 and Charles' was 0.81. John made 47 out of 86 free-throws. Who was the best free-throw shooter?
   a. Michael
   b. Larry
   c. Charles
   d. John

163. Which number below is described by the following statements? The hundredths digit is 4 and the tenths digit is twice the thousandths digit.
   a. 0.643
   b. 0.0844
   c. 0.446
   d. 0.0142

164. If a telephone pole weighs 11.5 pounds per foot, how much does a 32-foot pole weigh?
   a. 368 pounds
   b. 357 pounds
   c. 346 pounds
   d. 338.5 pounds

165. There are 5.802×10^{22} atoms in 8.23 grams of a certain molecule. Approximately how many grams does a single atom weigh?
   a. 7.05×10^{21}
   b. 4.78×10^{23}
   c. 1.42×10^{-22}
   d. 4.78×10^{-23}

166. Bill traveled 117 miles in 2.25 hours. What was his average speed?
   a. 26.3 miles per hour
   b. 5.2 miles per hour
   c. 46 miles per hour
   d. 52 miles per hour

167. Tom is cutting a piece of wood to make a shelf. He cut the wood to 3.5 feet, but it is too long to fit in the bookshelf he is making. He decides to cut 0.25 feet off the board. How long will the board be after he makes the cut?
   a. 3.25 feet
   b. 3.75 feet
   c. 3.025 feet
   d. 2.75 feet

168. A bricklayer estimates that he needs 6.5 bricks per square foot. He wants to lay a patio that will be 110 square feet. How many bricks will he need?
a. 650
b. 7,150
c. 6,500
d. 715

169. Find the area of a circle with a radius of 6 inches. The formula for the area of a circle is $A = \pi r^2$, where r is the radius. Use 3.14 for π.
a. 37.68 square inches
b. 113.04 square inches
c. 9.42 square inches
d. 75.36 square inches

170. Mary made 54 copies at the local office supply store. The copies cost $0.09 each. What was the total cost of the copies?
a. $4.86
b. $4.66
c. $3.86
d. $3.96

171. Tammi's new printer can print 13.5 pages per minute. How many pages can it print in 4 minutes?
a. 52
b. 48
c. 64
d. 54

172. The price of gasoline is $3.239 cents per gallon. If the price increases by three tenths of a cent, what will the price of gasoline be?
a. $3.539
b. $3.242
c. $3.269
d. $3.240

173. Louise is estimating the cost of the groceries in her cart. She rounds the cost of each item to the nearest dollar to make her calculations. If an item costs $1.45, to what amount will Louise round the item?
  a. $1.00
  b. $1.50
  c. $2.00
  d. $1.40

174. A pipe has a diameter of 2.5 inches. Insulation that is 0.5 inches thick is placed around the pipe. What is the diameter of the pipe with the insulation around it?
  a. 1.75 inches
  b. 4.5 inches
  c. 2.6 inches
  d. 3.5 inches

175. A bowling ball rolls down the lane at a speed of 3.7 feet per second. Assuming that 1 inch ≈ 2.54 cm, what is the speed of the bowling ball in centimeters per minute?
  a. 6,766.56 centimeters per minute
  b. 2,664 centimeters per minute
  c. 563.88 centimeters per minute
  d. 222 centimeters per minute

176. George worked from 7:00 A.M. to 3:30 P.M. with a 45-minute break. If George earns $10.50 per hour and does not get paid for his breaks, how much money did he earn? (Round to the nearest cent.)
  a. $89.25
  b. $81.40
  c. $97.13
  d. $81.38

177. Marci filled her car's gas tank on Monday, and the odometer read 32,461.3 miles. On Friday when the car's odometer read 32,659.7 miles, she filled the car's tank again. It took 12.4 gallons to fill the tank. How many miles to the gallon does Marci's car get?
   a. 16 miles per gallon
   b. 18.4 miles per gallon
   c. 21.3 miles per gallon
   d. 14 miles per gallon

178. Martha has $60 to spend and would like to buy as many calculators as possible with the money. The calculators that she wants to buy are $8.75 each. How much money will she have left over after she purchases the greatest possible number of calculators?
   a. $0.25
   b. $3.50
   c. $7.00
   d. $7.50

179. The distance from the Sun to Earth is approximately 9.3×10^7 miles. What is this distance, in miles, expressed in standard notation?
   a. 930,000,000
   b. 93,700,000
   c. 0.00000093
   d. 93,000,000

180. The distance from Earth to the moon is approximately 240,000 miles. What is this distance, in miles, expressed in scientific notation?
   a. 24×10^4
   b. 240×10^3
   c. 2.4×10^5
   d. 2.4×10^{-5}

181. Which of the following numbers will yield a number larger than 23.4 when it is multiplied by 23.4?
   a. 0.999
   b. 0.0008
   c. 0.3
   d. 1.0002

182. Kelly plans to fence in her yard. The Fabulous Fence Company charges $3.25 per foot of fencing and $15.75 an hour for labor. If Kelly needs 350 feet of fencing and the installers work a total of 6 hours installing the fence, how much will she owe the Fabulous Fence Company?
   a. $1,137.50
   b. $1,232.00
   c. $1,069.00
   d. $1,005.50

183. Thomas is keeping track of the rainfall in the month of May for his science project. The first day, 2.6 cm of rain fell. On the second day, 3.4 cm fell. On the third day, 2.1 cm fell. How many more cm of rain are needed to reach the average monthly rainfall in May, which is 9.7 cm?
   a. 8.1 cm
   b. 0.6 cm
   c. 1.6 cm
   d. 7.4 cm

184. Rick paddles his canoe 25.2 miles on the first day of his vacation at an average rate of 3.3 miles per hour. How long did his trip take? Round your answer to the nearest tenth of an hour.
   a. 7.2 hours
   b. 7.6 hours
   c. 8.3 hours
   d. 9.6 hours

185. Mona purchased one and a half pounds of turkey at the deli for $6.90. How much did she pay per pound?
   a. $4.60
   b. $10.35
   c. $3.45
   d. $5.70

186. Lucy is sending out flyers and pays a bulk rate of 14.9 cents per piece of mail. If she mails 1,500 flyers, what will she pay?
   a. $14.90
   b. $29.80
   c. $256.50
   d. $223.50

187. The length of one lap of a race track is 0.27 mile. A race car averages a speed of 110 miles per hour. How many laps does the car complete in 12 minutes on average? Round your answer to the nearest tenth of a lap.

    a. 22.0 laps

    b. 29.7 laps

    c. 81.5 laps

    d. 407.4 laps

Answer Explanations

126. **a.** Divide 2.3 × 10²¹ meters by 9.46 × 10¹⁵ meters per light year to get the number of light years:

$$(2.3 \times 10^{21}) \div (9.46 \times 10^{15}) \approx 243,128.9641$$

Since the tenths digit (the digit immediately following the decimal point) is greater than 5, increase the ones digit by 1 to round to the nearest light year. So, the star is approximately 243,129 light years from Earth.

127. **b.** The two items that Kara bought must be subtracted from the amount of money she brought with her: $26.00 – $6.27 – $14.98 = $4.75.

128. **a.** To find the profit, you must subtract what Lucas paid for the motorcycle from the sale price: $7,777.77 – $5,875.98 = $1,901.79.

129. **c.** If you add zeros to the end of Dan's and Ed's averages so that they all have three decimal places, it will be easy to compare the batting averages. The four averages are 0.349, 0.200, 0.350, and 0.299; 0.350 is the largest, which is Ed's batting average.

130. **d.** You must multiply the number of towels sold by the price of each towel: 85 × $6.95 = $590.75.

131. **d.** Call the original value of the computer x. The amount the computer is worth after one year is $0.85x$. Then, the next year, it depreciates again so that its value is $0.85(0.85x) = (0.85)^2x$. Likewise, after three years its value is 0.85 of its value at the end of the second year, which is $(0.85)^3x$, and finally its value after four years is $(0.85)^4x$. Since we are given that this value is $650, we have the equality $(0.85)^4x = 650$. To solve for x, divide both sides by $(0.85)^4$:

$$x = \frac{\$650}{(0.85)^4} \approx \$1,245.20.$$

132. **a.** To find the average number of miles Katie ran, you should divide the total number of miles by the number of days: 11.1 ÷ 3 = 3.7.

133. **d.** To find the price of each individual ticket, you should divide the total cost by the number of tickets purchased: $52.50 ÷ 6 = $8.75.

134. **b.** The value of the computer after one year is 0.85($2,200) = $1,870. Then, its value at the end of the second year is 0.85($1,870) = $1,589.50. And finally, at the end of the third year, its value is 0.85($1,589.50) = $1,351.08.

135. **c.** First, we need to determine the total number of minutes for which the call lasts. Since 1 hour = 60 minutes, the call lasts 68 minutes. The cost of the first 10 minutes is $0.50 per minute; this is $0.50(10) = $5.00. The cost for the remaining 58 minutes of the call is $0.22 per minute; this is $0.22(58) = $12.76. So, the total cost of the call is $5.00 + $12.76 = $17.76.

136. **b.** Find the difference between 0.24 and 0.25 mm by subtracting: 0.25 – 0.24 = 0.01 mm. Half of this is 0.01 ÷ 2 = 0.005. Add to 0.24 to get 0.245 mm.

137. **a.** Keri's three scores need to be added to find her total score. To add decimals, line up the numbers and decimal points vertically and add normally:

$$
\begin{array}{r}
5.65 \\
5.85 \\
+\ 5.75 \\
\hline
17.25
\end{array}
$$

138. **d.** To multiply decimals, multiply normally, count the number of decimal places in the problem, then use the same number of decimal places in the answer: 35 × $4.88 = $170.80. Since there are two decimal places in the problem, there should be two in the answer.

139. **a.** First, we need to determine the number of gallons of gas used for the trip. Divide the cost of the gas by the price per gallon: $256.66 ÷ $3.13 per gallon = 82 gallons. Next, divide the total number of miles traveled by the number of gallons of gas used: 1,703.6 ÷ 82 ≈ 20.78. So, they averaged approximately 21 miles per gallon of gas.

140. **c.** Convert both the width (5 in.) and length (7 in.) to centimeters by *multiplying* each by 2.54. The width is 5 × 2.54 = 12.70 cm, and the length is 7 × 2.54 = 17.78 cm. So, the dimensions in cm are 12.70 cm × 17.78 cm.

141. b. Last year's budget must be subtracted from this year's budget: $14.3 million – $12.4 million = $1.9 million. Since both numbers are *millions*, the 14.3 and 12.4 can simply be subtracted and *million* is added to the answer.

142. a. The three distances must be added together. To add decimals, line the numbers up vertically so that the decimal points are aligned. Then, add normally:

$$
\begin{array}{r}
3.1 \\
4.25 \\
+ \ 10.8 \\
\hline
18.15
\end{array}
$$

143. d. The fastest time is the smallest number. If you chose **c**, you chose the slowest time since it is the largest number (this person took the longest amount of time to finish the race). To compare decimals easily, add zeros at the end so that all numbers have the same number of decimal places: $10.09 < 10.10 < 10.14 < 10.20$. (Note: adding zeros to the end of a number, to the right of the decimal point, does not change the value of the number.)

144. d. The number of *hours* the man sleeps in one 365-day year equals $365 \times 9 = 3{,}285$ hours. To find the number of *minutes* to which this is equal, multiply by 60 minutes per hour: $3{,}285 \times 60 = 197{,}100$ minutes. Finally, to find the number of minutes he has slept so far in 38 years, multiply the minutes per year by 38 years: $197{,}100 \times 38 = 7{,}489{,}800$ minutes.

145. d. To compare decimals, you can add zeros to the end of the number after the decimal point (this will not change the value of the number): $0.315 < 0.317 < 0.320$. Choice **a** is incorrect because 0.311 is smaller than 0.315. Choice **b** is incorrect because 0.309 is smaller than 0.315. Choice **c** is incorrect because 0.321 is larger than 0.32.

146. b. First, determine the approximate number of professors (call this number x) that currently work at the school by setting up a proportion of the form $\frac{students}{professors}: \frac{1{,}450}{x} = \frac{16}{1}$. Solve for x by cross-multiplying and then dividing both sides by 16: $16x = 1{,}450$, so $x = 90.625$. Since we cannot have a fraction of a professor, round up to 91. Next, to determine the approximate number of professors (call this number y) that *would* be working at the school if the ratio were 1:12, set up a similar proportion of the form $\frac{students}{professors}: \frac{1,450}{y} = \frac{12}{1}$. Solve for y by cross-multiplying and then dividing both sides by 12: $12y = 1,450$, so that $y \approx 120.83$. Rounding up, we see that 121 professors would yield such a ratio. Therefore, to find the number of additional professors that would need to be hired in order to yield this ratio, subtract $121 - 91$ to get 30.

147. **a.** The school record is less than Brian's time. Therefore, 2.68 must be subtracted from 13.4. To subtract decimals, line up the numbers vertically so that the decimal points are aligned.

$$\begin{array}{r} 13.40 \\ -\ 2.68 \\ \hline 10.72 \end{array}$$

Since 13.4 has one fewer decimal place than 2.68, you must add a zero after the 4 (13.40) before subtracting. After you have done this, subtract normally. If you chose **d**, you added instead of subtracted.

148. **d.** To find each installment, the total yearly cost ($390) must be divided by the number of payments (12): $\$390 \div 12 = \$32.50$. Choices **a** and **c** do not make sense because they would mean that each monthly installment is more than the total yearly cost.

149. **c.** To find out how much greater one number is than another, you need to subtract. To subtract decimals, line the numbers up vertically so that the decimal points align. Then, subtract normally.

$$\begin{array}{r} 0.0101 \\ -\ 0.0030 \\ \hline .0071 \end{array}$$

150. **d.** First, we need to determine the number of ounces that are in $2\frac{1}{2}$ pounds. Since one pound equals 16 ounces, the number of ounces in $2\frac{1}{2}$ pounds equals $2\frac{1}{2} \times 16 = \frac{5}{2} \times 16 = 40$. Next, to determine the cost of 40 ounces of turkey breast (call this x), we set up a proportion of the form $\frac{ounces}{dollars}: \frac{9\ ounces}{\$4.80} = \frac{40\ ounces}{x}$. To solve for x, cross-multiply and then divide both sides by 9: $9x = 40(\$4.80) = \$192$, so $x \approx \$21.33$.

151. d. To find how much Jay spent, you must multiply the cost of each stamp ($0.44) by the number of stamps purchased (25): $0.44 × 25 = $11.00. To multiply decimals, multiply normally, then count the number of decimal places in the problem. Place the decimal point in the answer so that it contains the same number of decimal places as the problem does.

152. d. The cost of one gallon of gas at the end of the first week is $2.80 + 0.06 × ($2.80) ≈ $2.97. Then, the cost of one gallon of gas at the end of the second week is $2.97 + 0.06 × ($2.97) ≈ $3.15.

153. d. You must find the difference (subtraction) between Hanna's goal and what she has already sold. Add a decimal and two zeros to the end of $5,000 ($5,000.00) to make the subtraction easier: $5,000.00 − $3,574.38 = $1,425.62.

154. c. Let x represent the number of inches that represents 473.8 miles. Set up a proportion of the form $\frac{inches}{miles}$: $\frac{1.5 \text{ in.}}{62.4 \text{ mi.}} = \frac{x}{473.8 \text{ mi}}$. To solve for x, cross-multiply and then divide both sides by 62.4: $62.4x = (1.5)(473.8) = 710.7$, so $x ≈ 11.389 ≈ 11.4$ in.

155. b. To find the average, you must add the items ($97 + 78 + 84 + 86 = 345$) and divide the sum by the total number of items (4): $345 ÷ 4 = 86.25$.

156. c. You must multiply 11.7 feet per second by 5 seconds: $11.7 × 5 = 58.5$ feet. To multiply decimals, multiply normally, then count the total number of decimal places in the problem and move the decimal point in the answer so that it contains the same number of decimal places. If you choose **a**, you forgot to add the decimal point after you multiplied.

157. a. First, determine the total amount that you spent in buying 250 shares at $23.81 per share by multiplying: $250 × ($23.81) = $5,952.50. Next, similarly determine the total amount your investment of 250 shares is worth at the current price of $19.28 per share by multiplying: $250 × ($19.28) = $4,820.00. Finally, the loss you would incur is obtained by subtracting these two totals: $5,952.50 − $4,820.00 = $1,132.50.

158. a. One hour and thirty minutes is $1\frac{1}{2}$ hours or 1.5 hours. Therefore, multiply the number of miles Mike can jog in one hour by 1.5 to find the number he can jog in an hour and a half: $6.5 × 1.5 = 9.75$ miles.

159. a. The hash marks indicate units of 0.01 between 0.75 and 0.80. Point A is 0.77. See the figure below.

160. c. Nicole has 15 pounds of chocolate to divide into 20 baskets. Divide 15 by 20: $15 \div 20 = 0.75$ pounds per basket.

161. b. Quickly compare decimals by adding zeros to the end of a decimal so that all numbers being compared have the same number of decimal places.

Choice **a** does *not* work:
635.80
635.90
635.93—the book's call number

Choice **b** does work:
635.80
635.930—the book's call number
635.95

Choice **c** does *not* work:
635.930—the book's call number
635.935
635.940

Choice **d** does *not* work:
635.93—the book's call number
635.99
636.00

162. **c.** Change all of the comparisons to decimals by dividing the number of free-throws made by the number attempted. Michael's average is 19 ÷ 30 = 0.633, John's is 0.546, Larry's is given as 0.745, and Charles' is given as 0.81. The person with the largest decimal average is the best free-throw shooter. Add zeros to the ends of the decimals to compare easily. The shooters are listed from best to worst below.
0.810 Charles
0.745 Larry
0.633 Michael
0.546 John

163. **a.** From left to right, the first decimal place is the tenths, the second is the hundredths, and the third is the thousandths. The first criterion is that the hundredths digit is 4. The second decimal place is 4 only in choice **a** and choice **c**. The second criterion is that the first decimal place is twice the third decimal place. This is only true in choice **a**, in which 6 is twice 3.

164. **a.** Multiply 11.5 pounds per foot by 32 feet: $11.5 \times 32 = 368$ pounds.

165. **c.** Let x be the number of grams that a single atom in the molecule weighs. Set up a proportion of the form $\frac{atoms}{grams} : \frac{5.802 \times 10^{22}\ atoms}{8.23\ grams} = \frac{1\ atom}{x}$. To solve for x, cross-multiply and then divide both sides by 5.802×10^{22}:
$(5.802 \times 10^{22})\,x = 8.23$, so $x = \frac{8.23}{5.802 \times 10^{22}} \approx 1.42 \times 10^{-22}$ grams.

166. **d.** Use the formula $d = rt$ (distance = rate × time). Substitute 117 miles for d, substitute 2.25 hours for t, and solve for r.
$117 = 2.25r$
$\frac{117}{2.25} = \frac{2.25r}{2.25}$
$r = 52$
Bill's average speed was 52 miles per hour.

167. **a.** Subtract 0.25 from 3.5: $3.5 - 0.25 = 3.25$ feet.

168. **d.** Multiply 6.5 bricks per square foot by 110 square feet: $6.5 \times 110 = 715$ bricks.

169. b. Substitute 6 for r in the formula $A = \pi r^2$ and solve for A.

$A = (3.14)(6)^2$

$A = (3.14)(36)$

$A = 113.04$

The area of the circle is 113.04 square inches.

A common mistake in this problem is to say that 6^2 is 12. This is NOT true; 6^2 means 6×6, which equals 36.

170. a. Multiply 54 copies by $0.09 per copy to find the total cost: $54 \times \$0.09 = \$4.86$.

171. d. Multiply 13.5 pages per minute by 4 minutes to find the number of copies made: $13.5 \times 4 = 54$ copies.

172. b. Three tenths of a cent can be written as 0.3¢, or changed to dollars by moving the decimal point two places to the left, $0.003. If $0.003 is added to $3.239, the answer is $3.242.

173. a. $1.45 rounded to the nearest dollar is $1.00. You are rounding to the ones place, so look at the place to the right (the tenths place) to decide whether to round up or stay the same. Since 4 is less than 5, the 1 stays the same and the places after the 1 become zero.

174. d. The insulation surrounds the whole pipe. If the diameter of the pipe is 2.5 inches, the insulation will add 0.5 inches on both sides of the diameter. See the diagram below: $2.5 + 0.5 + 0.5 = 3.5$ inches.

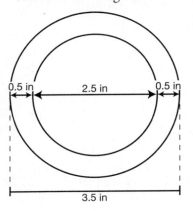

175. **a.** Use the facts that 1 minute = 60 seconds, 1 foot = 12 inches, and 1 inch ≈ 2.54 centimeters to convert the units, as follows:

3.7 feet	60 seconds	12 inches	2.54 centimeters
1 second	1 minute	1 foot	1 inch

Multiply the numbers across the top and bottom, and cancel pairs of like units in the top and bottom to conclude that 3.7 feet per second = (3.7 × 60 × 12 × 2.54) centimeters per minute.

Simplifying yields 6,766.56 centimeters per minute.

176. **d.** First, find the number of hours George worked. From 7:00 A.M. to 3:30 P.M. is $8\frac{1}{2}$ hours. Take away his $\frac{3}{4}$ hour break and he works $7\frac{3}{4}$ hours. To find what George is paid, multiply the hours worked, 7.75 (changed from the fraction), by the pay per hour, $10.50: 7.75 × $10.50 = $81.375. The directions say to round to the nearest cent. Therefore, the answer is $81.38.

177. **a.** Since the tank was full on Monday, whatever it takes to fill the tank is the amount of gas that she has used. Therefore, she has used 12.4 gallons of gas. Next, find the number of miles Marci traveled by subtracting Monday's odometer reading from Friday's odometer reading: 32,659.7 − 32,461.3 = 198.4 miles. Divide the miles driven by the gas used to find the miles per gallon: 198.4 ÷ 12.4 = 16 miles per gallon.

178. **d.** Divide the $60 by $8.75 to find the number of calculators Martha can buy: $60 ÷ $8.75 = 6.857. She can buy 6 calculators. She doesn't have enough to buy a seventh calculator. This means that she has spent $52.50 on calculators because $8.75 × 6 = $52.50. To find how much money she has left, subtract $52.50 from $60. The answer is $7.50.

179. **d.** In order to convert this number to standard notation, multiply 9.3 by the factor of 10^7. Since 10^7 is equal to 10,000,000, 9.3 × 10,000,000 is equal to 93,000,000. As an equivalent solution, move the decimal point in 9.3 seven places to the right since the exponent on the 10 is positive 7.

180. c. To convert to scientific notation, place a decimal point after the first non-zero digit to create a number between 1 and 10—in this case, between the 2 and the 4. Count the number of decimal places from that decimal to the place of the decimal in the original number. In this case, the number of places would be 5. This number, 5, becomes the exponent of 10 and is positive because the original number was greater than one. The answer then is 2.4×10^5.

181. d. When multiplying by a number less than 1, you get a product that is less than the number you started with. Multiplying by a number greater than 1 gives you a larger number than the one you started with. Therefore, multiplying by 1.0002 will yield a number larger than 23.4.

182. b. First, figure out the cost of the fence by multiplying the number of feet of fence by $3.25 per foot: $350 \times \$3.25 = \$1,137.50$. Next, find the cost of the labor by multiplying the hours of labor by $15.75 per hour: $6 \times \$15.75 = \$94.50$. Add the two costs together to find how much Kelly owes Fabulous Fence: $\$1,137.50 + \$94.50 = \$1,232.00$.

183. c. Find the total amount of rain that has fallen so far in May: $2.6 + 3.4 + 2.1 = 8.1$ cm. Find the difference between this amount and the average rainfall for May by subtracting: $9.7 - 8.1 = 1.6$ cm.

184. b. Using *distance = rate × time*, the approximate length of this trip is 25.2 miles ÷ 3.3 miles per hour ≈ 7.6 hours.

185. a. Divide the cost of the turkey by the weight: $\$6.90 \div 1.5 = \$4.60$ per pound.

186. d. Multiply the price per piece of mail by the number of pieces: $14.9 \times 1,500 = 22,350$ cents. Change the cents into dollars by dividing by 100 (move the decimal point two places to the left): 22,350 cents = $223.50.

187. c. First, we must determine the number of miles (call this x) traveled in 12 minutes. To do so, set up a proportion of the form $\frac{miles}{minutes} : \frac{110\ miles}{60\ minutes} = \frac{x}{12\ minutes}$. To solve for x, cross multiply and then, divide both sides by 60: $60x = 110(12) = 1,320$, so $x = 22$ miles. Next, we must determine the number of laps (call this y) corresponding to 22 miles. To do so, we set up another proportion, this time of the form $\frac{laps}{miles} : \frac{1\ lap}{0.27\ miles} = \frac{y}{22\ miles}$. To solve for y, cross multiply and then divide both sides by 0.27: $0.27y = 22(1) = 22$, so $y = 22 \div 0.27 \approx 81.5$ laps.

Percents

Percentages have many everyday uses, from figuring out the tip in a restaurant to understanding interest rates. This chapter will give you practice in solving word problems that involve percents.

188. A pair of pants costs $24. If the cost is reduced by 8%, what is the new cost of the pants?
   a. $25.92
   b. $21.06
   c. $22.08
   d. $16.00

189. Michael scored 260 points during his junior year on the school basketball team. He scored 25% more points during his senior year. How many points did he score during his senior year?
   a. 195
   b. 65
   c. 325
   d. 345

190. Brian is a real estate agent. He earns a 2.5% commission on each sale. During the month of June he sold three houses. The houses sold for $153,000, $399,000, and $221,000. What was Brian's total commission on these three sales?
   a. $193,250
   b. $11,460
   c. $3,825
   d. $19,325

191. Cory purchased a frying pan that was on sale for 30% off. She saved $3.75 with the sale. What was the original price of the frying pan?
   a. $10.90
   b. $9.25
   c. $12.50
   d. $11.25

192. Peter purchased 14 new baseball cards for his collection. This increased the size of his collection by 35%. How many baseball cards does Peter now have?
   a. 5
   b. 54
   c. 40
   d. 34

193. Joey has 30 pages to read for history class tonight. He decided that he would take a break when he finished reading 70% of the pages assigned. How many pages must he read before he takes a break?
   a. 7
   b. 21
   c. 9
   d. 18

194. An investor earned $313.62 in interest in one year on a mutual fund account that paid 3.85% simple interest. How much more interest would the account have earned if the interest rate had been 4.35%?
   a. $40.73
   b. $354.35
   c. $2,126.10
   d. $8,145.97

195. Nick paid $68.25 for a coat, the price of which includes a sales tax of 5%. What was the original price of the coat before tax?
a. $63.25
b. $65.25
c. $65.00
d. $64.84

196. The Dow Jones Industrial Average fell 2% today. The Dow began the day at 10,600. What was the Dow at the end of the day after the 2% drop?
a. 10,400
b. 10,812
c. 10,388
d. 7,800

197. The population of Hamden was 350,000 in 1990. By 2000, the population had decreased to 329,000. What percent of decrease is this?
a. 16%
b. 60%
c. 6%
d. 6.4%

198. Connecticut state sales tax is 6%. Lucy purchases a picture frame that costs $10.50. What is the Connecticut sales tax on this item?
a. $0.60
b. $6.30
c. $0.63
d. $1.05

199. Wendy brought $15 to the mall. She spent $6 on lunch. What percent of her money did she spend on lunch?
a. 60%
b. 40%
c. 4%
d. 33.3%

200. Assuming the total surface area of Earth is approximately 5.2×10^{14} square meters, and that the continents, all taken together, comprise about 30% of the total surface area, how many square meters of the Earth's surface are covered by land?

    a. 7.8×10^{13} square meters

    b. 1.56×10^{14} square meters

    c. 2.34×10^{14} square meters

    d. 3.64×10^{14} square meters

201. Kara borrowed $3,650 for one year at an annual interest rate of 16%. How much did Kara pay in interest?

    a. $1,168.00

    b. $584.00

    c. $4,234.00

    d. $168.00

202. Rebecca is 12.5% taller than Debbie. Debbie is 64 inches tall. How tall is Rebecca?

    a. 42 inches

    b. 8 inches

    c. 56 inches

    d. 72 inches

203. Kyra receives a 5% commission on every car she sells. She received a $1,325 commission on the last car she sold. What was the cost of the car?

    a. $26,500.00

    b. $66.25

    c. $27,825.00

    d. $16,250.00

204. A tent originally sold for $260 and has been marked down to $208. What is the percent of the discount?

    a. 20%

    b. 25%

    c. 52%

    d. 18%

205. The football boosters club had 80 T-shirts made to sell at football games. By mid-October, they had only 12 left. What percent of the shirts had been sold?
   a. 85%
   b. 15%
   c. 60%
   d. 40%

206. A printer that sells for $190 is on sale for 15% off. What is the sale price of the printer?
   a. $161.50
   b. $175.00
   c. $140.50
   d. $156.50

207. Consider the following two quantities:
   I. A 15% discount followed by a 15% mark-up on an item originally priced at x dollars.
   II. A 15% mark-up followed by a 15% discount on an item originally priced at x dollars.
   Which of the following statements is true?
   a. **I** is greater than **II**
   b. **I** is less than **II**
   c. **I** equals **II**
   d. It is unable to make comparison of **I** and **II**.

208. There are 81 female teachers at Russell High. If 45% of the teachers in the school are female, how many teachers are there at Russell High?
   a. 180
   b. 36
   c. 165
   d. 205

209. Kim is a medical supplies salesperson. Each month she receives a 5% commission on all her sales of medical supplies up to $20,000 and 8.5% on her total sales over $20,000. Her total commission for May was $3,975. What were her sales for the month of May?
   a. $79,500
   b. $35,000
   c. $65,500
   d. $55,000

210. Fill in the blank: If Tanya has 39% as many CDs as Matt, and Joan has 42% as many CDs as Tanya, then Joan has _____ as many CDs as Matt.
   a. 3.5%
   b. 16.38%
   c. 45%
   d. 81%

211. Christie purchased a scarf marked $15.50 and gloves marked $5.50. Both items were on sale for 20% off the marked price. Christie paid 5% sales tax on her purchase. How much did she spend?
   a. $25.20
   b. $16.80
   c. $26.46
   d. $17.64

212. Last year, a math textbook cost $54. This year the cost is 107% of what it was last year. How much does the textbook cost this year?
   a. $59.78
   b. $57.78
   c. $61.00
   d. $50.22

213. Larry earned $32,000 per year. He then received a $3\frac{1}{4}$% raise. What is Larry's salary after the raise?
   a. $33,040
   b. $35,000
   c. $32,140
   d. $32,960

214. Bill spent 50% of his savings on school supplies, then he spent 50% of what was left on lunch. If he had $8 left after lunch, how much did he have in savings at the beginning?
  a. $32
  b. $16
  c. $30
  d. $18

215. The enrollment in yoga classes is calculated at the end of each month. It was found that enrollments increased by 3.7% during January, then decreased by 1.7% during February, and then increased by 0.8% in March. What was the net percent change in enrollments over this three-month period? Round your answer to the nearest hundredth of a percent.
  a. 0.0275%
  b. 1.93%
  c. 2.75%
  d. 3.15%

216. Kristen earns $550 each week after taxes. She deposits 10% of her income into a savings account and 7% into a retirement fund. How much does Kristen have left after the money is taken out for her savings account and retirement fund?
  a. $505.25
  b. $435.50
  c. $533.00
  d. $456.50

217. Coastal Cable had 1,440,000 customers in January of 2002. During the first half of 2002 the company launched a huge advertising campaign. By the end of 2002 the company had 1,800,000 customers. What is the percent of increase?
  a. 36%
  b. 21%
  c. 20%
  d. 25%

218. The price of heating oil rose from $1.70 per gallon to $2.38 per gallon. What is the percent of increase?

 a. 40%

 b. 33%

 c. 29%

 d. 38%

219. 450 girls were surveyed about their favorite sport: 24% said that basketball is their favorite sport, 13% said that ice hockey is their favorite sport, and 41% said that softball is their favorite sport. The remaining girls said that field hockey is their favorite sport. What percent of the girls surveyed said that field hockey is their favorite sport?

 a. 37%

 b. 22%

 c. 78%

 d. 35%

220. At Yale New Haven Hospital, 25% of babies born weigh less than 6 pounds and 78% weigh less than 8.5 pounds. What percent of the babies born at Yale New Haven Hospital weigh between 6 and 8.5 pounds?

 a. 22%

 b. 24%

 c. 53%

 d. 2.5%

221. An $80.00 coat is marked down 20%. It does not sell, so the shop owner marks it down an additional 15%. What is the new price of the coat?

 a. $64.00

 b. $68.60

 c. $52.00

 d. $54.40

222. The price of a suit decreased from $250 to $210. Later, the price was further reduced from $210 to $170. Which of the following statements is true?
    a. The percentage decrease for the first markdown is larger than the percentage decrease for the second.
    b. The percentage decrease for the second markdown is larger than the percentage decrease for the first.
    c. The percentage decrease for the first markdown is equal to the percentage decrease for the second.
    d. The percentage decreases cannot be computed using the given information.

223. An exam has 25 questions. Sue earned a 76% on the exam. How many questions did she get correct?
    a. 18
    b. 19
    c. 20
    d. 21

224. In Kimmi's fourth grade class, 8 out of the 20 students walk to school. What percent of the students in her class walk to school?
    a. 40%
    b. 50%
    c. 45%
    d. 35%

225. In Daniel's fifth grade class, 37.5% of the 24 students walk to school. One-third of the walkers got a ride to school today from their parents. How many walkers got a ride to school from their parents today?
    a. 9
    b. 12
    c. 2
    d. 3

226. Lindsay purchased a pocketbook for $35, a pair of shoes for $45, and a T-shirt for $20. The sales tax on the items was 6%. How much sales tax did she pay?
  a. $2.70
  b. $3.30
  c. $6.00
  d. $6.60

227. Wendy bought a book, and the sales tax on the book was $2.12. If the sales tax is 8%, what was the price of the book?
  a. $26.50
  b. $16.96
  c. $24.76
  d. $265.00

228. Mr. Pelicas took his family out to dinner. The bill was $65.00. He would like to leave a 17% tip. How much should he leave?
  a. $17.00
  b. $3.25
  c. $11.05
  d. $16.25

229. The *Daily News* reported that 54% of people surveyed said that they would vote for Larry Salva for mayor. Based on the survey results, if 23,500 people vote in the election, how many people are expected to vote for Mr. Salva?
  a. 12,690
  b. 4,350
  c. 10,810
  d. 18,100

230. Bikes are on sale for 30% off the original price. What percent of the original price will the customer pay if he gets the bike at the sale price?
  a. 130%
  b. 60%
  c. 70%
  d. 97%

231. Forty percent of the adults attending a family reunion brought a soup or salad. If 10 adults brought a soup or salad, how many adults attended the reunion?

   a. 25
   b. 30
   c. 40
   d. 50

232. In response to the threat of a winter storm, the local supply store marks up the price of snow shovels by 250%. If the snow shovel now sells for $40, what was its price prior to threat of the storm?

   a. $100
   b. $20
   c. $16
   d. $10

233. The freshman class is participating in a fundraiser. Their goal is to raise $5,000. After the first two days of the fundraiser, they had raised 32% of their goal. How much money did they raise during the first two days?

   a. $160
   b. $32
   c. $1,600
   d. $3,400

234. A recipe for a large kettle of yams calls for $2\frac{3}{8}$ cups of brown sugar. Melinda has only $\frac{3}{4}$ cup of brown sugar. What percentage of the recipe can she make? Round your answer to the nearest percent.

   a. 178%
   b. 65%
   c. 40%
   d. 32%

235. Peter was 60 inches tall on his thirteenth birthday. By the time he turned 15, his height had increased by 15%. How tall was Peter when he turned 15?

   a. 75 inches
   b. 69 inches
   c. 72 inches
   d. 71 inches

236. Laura paid $80 for a pair of jeans. The ticketed price was 20% off the original price, plus the sign on the rack said, "Take an additional 15% off the ticketed price." What was the original price of the jeans, to the nearest cent?
a. $96.25
b. $105.50
c. $117.65
d. $120.00

237. The value of a baseball card increased by 325% during the past year. If the value of the card is now $520, what was its approximate value last year?
a. $1,690
b. $1,040
c. $160
d. $130

238. What percent of the figure below is shaded?

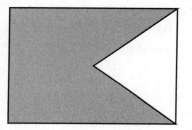

a. 50%
b. 65%
c. 75%
d. 80%

239. Melisa and Jennifer threw a fiftieth birthday party for their father at a local restaurant. When the bill came, Melisa added a 15% tip of $42. Jennifer said that the service was wonderful and they should leave a 20% tip instead. How much is a 20% tip?
a. $56
b. $45
c. $47
d. $60

240. The Hamden Town Manager wants to know what percent of the snow removal budget has already been spent. The budget for snow removal is $130,000. It has been an exceptionally snowy year, and the town has already spent $100,000 for snow removal. What percent of the budget has already been spent? Round to the nearest percent.
   a. 30%
   b. 70%
   c. 72%
   d. 77%

241. A real estate agent earns a 1.5% commission on her sales. What is her commission if she sells a $359,000 house?
   a. $53,850
   b. $5,385
   c. $23,933
   d. $1,500

242. The manager of a specialty store marks up imported products 110%. If a vase imported from Italy costs him $42, what price tag will he put on the item?
   a. $84.00
   b. $83.50
   c. $65.00
   d. $88.20

243. Bob's year-end bonus is equal to $1\frac{3}{4}$% of his salary. If the bonus is $800, what was his salary for the year? Round your answer to the nearest dollar.
   a. $45,714
   b. $60,526
   c. $75,438
   d. $80,000

244. Kyra's weekly wages are $895. A Social Security tax of 7.51% and a State Disability Insurance of 1.2% are taken out of her wages. How much is her weekly paycheck, assuming there are no other deductions?
   a. $827.79
   b. $884.26
   c. $962.21
   d. $817.05

245. Oscar's Oil Company gives customers a 5% discount if they pay their bill within 10 days. The Stevens' oil bill is $178. How much will they save if they pay the bill within 10 days?
   a. $8.90
   b. $5.00
   c. $17.80
   d. $14.60

246. Josephine is on an 1,800 calorie per day diet. She tries to keep her intake of fat to no more than 30% of her total calories. Based on an 1,800 calorie a day diet, what is the maximum number of calories that Josephine should consume from fats per day to stay within her goal?
   a. 600
   b. 640
   c. 580
   d. 540

247. A family may deduct 24% of their childcare expenses from their income tax owed. If a family had $1,345 in childcare expenses, how much can they deduct?
   a. $1,022.20
   b. $345.00
   c. $322.80
   d. $789.70

248. As a New Year's resolution, Mark decided to decrease his coffee consumption. Each day, he vowed to drink 95% of the amount of coffee he drank the previous day. If he drinks five cups of coffee on Day 1, how many cups of coffee will he drink on Day 7? Round your answer to the nearest tenth of a cup.
   a. 3.5 cups
   b. 3.7 cups
   c. 3.9 cups
   d. 4.1 cups

249. Sales increased by only $\frac{1}{2}$% last month. If the sales from the previous month were $152,850, what were last month's sales?

    a. $229,275.00

    b. $153,614.25

    c. $152,849.05

    d. $151,397.92

250. Laura is planning her wedding. She expects 220 people to attend the wedding, but she has been told that approximately 5% typically don't show. About how many people should she expect not to show?

    a. 46

    b. 5

    c. 11

    d. 23

Answer Explanations

188. c. If the cost of the pants is reduced by 8%, the cost of the pants is 92% of the original cost (100% − 8% = 92%). To find 92% of the original cost, multiply the original cost of the pants by the decimal equivalent of 92%: $24 × 0.92 = $22.08.

189. c. If the number of points Michael scored is increased by 25%, the number of points scored in his senior year is 125% of the number of points scored in his junior year (100% + 25% = 125%). To find 125% of the number of points scored in his junior year, multiply the junior year points by the decimal equivalent of 125%: 260 × 1.25 = 325. If you chose **a**, you calculated what his points would be if he scored 25% *fewer* points in his senior year than he did in his junior year.

190. d. First, find the total of Brian's sales: $153,000 + $399,000 + $221,000 = $773,000. To find 2.5% of $773,000, multiply by the decimal equivalent of 2.5%: $773,000 × 0.025 = $19,325. If you chose **a**, you used the decimal 0.25, which is 25%, not 2.5%.

191. c. Use a proportion to find the original cost of the frying pan: $\frac{part}{whole} = \frac{\%}{100}$. The $3.75 that was saved is *part* of the original price. The *whole* price is what we are looking for, so call it x. The % is 30 (the percent off): $\frac{3.75}{x} = \frac{30}{100}$. To solve the proportion, cross-multiply: $(3.75)(100) = 30x$. Divide both sides by 30 to solve for x: $\frac{375}{30} = \frac{30x}{30}$; $x = 12.50.

192. b. First, you must find how many baseball cards Peter had originally. Use a proportion to find the original number of baseball cards: $\frac{part}{whole} = \frac{\%}{100}$. The 14 baseball cards that he added to his collection is the *part*. The *whole* number of baseball cards is what we are looking for, so call it x. The % is 35 (the percent of increase): $\frac{14}{x} = \frac{35}{100}$. To solve the proportion, cross-multiply: $(14)(100) = 35x$. Divide both sides by 35 to solve for x: $\frac{1,400}{35} = \frac{35x}{35}$; $x = 40$. The original number of baseball cards was 40, and 14 more were added to the collection for a total of 54 cards.

193. b. To find 70% of 30, you must multiply 30 by the decimal equivalent of 70%: 30 × 0.70 = 21. If you chose **c**, you calculated how many pages he has left to read after his break.

194. **a.** Let x be the amount invested in the account. The interest earned is computed by multiplying x by the interest rate, 0.0385. This yields the equation $0.0385x = \$313.62$. To solve for x, divide both sides by 0.0385; $x \approx \$8,145.97$. Now, if the interest rate were instead 4.35%, then the amount of interest earned on $8,145.97 would be $8,145.97 \times 0.0435 \approx$ $354.35. Therefore, the investor would have earned $354.35 − $313.62 = \$40.73 *more* at this higher interest rate.

195. **c.** Since 5% sales tax was added to the cost of the coat, $68.25 is 105% of the original price of the coat. Use a proportion to find the original cost of the coat: $\frac{part}{whole} = \frac{\%}{100}$. *Part* is the price of the coat with the sales tax, $68.25. *Whole* is the original price on the coat that we are looking for. Call it x. The % is 105: $\frac{68.25}{x} = \frac{105}{100}$. To solve for x, cross-multiply: $(68.25)(100) = 105x$. Divide both sides by 105: $\frac{6,825}{105} = \frac{105x}{105}$; $x = \$65.00$.

196. **c.** The Dow lost 2%, so at the end of the day it is worth 98% of what it was worth at the beginning of the day (100% − 2% = 98%). To find 98% of 10,600, multiply 10,600 by the decimal equivalent of 98%: $10,600 \times 0.98 = 10,388$.

197. **c.** First, find the number of residents who left Hamden by subtracting the new population from the old population: 350,000 − 329,000 = 21,000. The population decreased by 21,000. To find what percent this is of the original population, divide 21,000 by the original population of 350,000: $21,000 \div 350,000 = 0.06$; 0.06 is equivalent to 6%. If you chose **d**, you found the decrease in relation to the NEW population (2000) when the decrease must be in relation to the original population (1990).

198. **c.** Find 6% of $10.50 by multiplying $10.50 by the decimal equivalent of 6%): $10.50 \times 0.06 = \$0.63$. If you chose **b**, you found 60% (0.6) instead of 6% (0.06).

199. **b.** Divide $6 by $15 to find the percent: $6 \div \$15 = 0.40$; 0.40 is equivalent to 40%.

200. **b.** The amount of surface area corresponding to land is 30% of 5.2×10^{14} square meters. This is computed by the product $0.30 \times (5.2 \times 10^{14}) = 1.56 \times 10^{14}$ square meters.

201. **b.** To find 16% of $3,650, multiply $3,650 by the decimal equivalent of 16: $3,650 × 0.16 = $584.00.

202. **d.** Since Rebecca is 12.5% taller than Debbie, she is 112.5% of Debbie's height (100% + 12.5% = 112.5%). To find 112.5% of Debbie's height, multiply Debbie's height by the decimal equivalent of 112.5%: 64 × 1.125 = 72 inches. If you chose **c**, you found what Rebecca's height would be if she were 12.5% *shorter* than Debbie (you subtracted instead of added).

203. **a.** Use the proportion $\frac{part}{whole} = \frac{\%}{100}$ to solve the problem; $1,325 is the *part* and 5% is the %. We are looking for the *whole*, so we will call it x: $\frac{1,325}{x} = \frac{5}{100}$. Cross multiply: $(1,325)(100) = 5x$. Divide both sides by 5 to solve for x: $\frac{132,500}{5} = \frac{5x}{5}$; $x = $26,500$. If you chose **b**, you found 5% of her commission (5% of $1,325).

204. **a.** Find the number of dollars off: $260 – $208 = $52. Next, determine what percent of the original price $52 is by dividing $52 by the original price, $260: $52 ÷ $260 = 0.20; 0.20 is equivalent to 20%.

205. **a.** Determine the number of T-shirts sold: 80 – 12 = 68. To find what percent of the original number of shirts 68 is, divide 68 by 80: 68 ÷ 80 = 0.85; 0.85 is equivalent to 85%. If you chose **b**, you found the percent of T-shirts that were *left* instead of the percent that had been *sold*.

206. **a.** The printer is 15% off, which means that it is 85% of its original price (100% – 15% = 85%). To find 85% of $190, multiply $190 by the decimal equivalent of 85%: $190 × 0.85 = $161.50.

207. **b.** <u>Quantity **I** is computed as follows</u>: Applying a 15% discount to x dollars yields the new price $x - 0.15x = 0.85x$. Then, applying a 15% mark-up on this price yields the new price $0.85x + 0.15(0.85x) = 0.85x + 0.1275x = 0.9775x$.
<u>Quantity **II** is computed as follows</u>: Applying a 15% mark-up to x dollars yields the new price $x + 0.15x = 1.15x$. Then, applying a 15% discount on this price yields the new price $1.15 - 0.15(1.15x) - 0.1725x = 0.9775x$. So, we see that Quantities **I** and **II** are equal.

208. **a.** Use the proportion $\frac{part}{whole} = \frac{\%}{100}$. *Part* is the number of female teachers (81). *Whole* is what we are looking for, call it x, and the % is 45: $\frac{81}{x} = \frac{45}{100}$. Cross multiply: $(81)(100) = 45x$. Divide both sides by 45 to solve for x: $\frac{8,100}{45} = \frac{45x}{45}$: $x = 180$ teachers.

209. **d.** Kim sold over \$20,000 in May. She received a 5% commission on the first \$20,000 of sales. To find 5%, multiply by the decimal equivalent of 5%: \$20,000 × 0.05 = \$1,000. Since her total commission was \$3,975, \$3,975 – \$1,000 = \$2,975 is the amount of commission she earned on her sales over \$20,000. \$2,975 represents 8.5% of her sales over \$20,000. To find the amount of her sales over \$20,000, use a proportion: $\frac{part}{whole} = \frac{\%}{100}$. *Part* is \$2,975, and *whole* is what we are looking for, so let's call it x. The % is 8.5: $\frac{2,975}{x} = \frac{8.5}{100}$. To solve for x, cross multiply: (2,975)(100) = 8.5x. Divide both sides by 8.5 to solve: $\frac{297,50}{8.5} = \frac{8.5x}{8.5}$; x = \$35,000. Her sales over \$20,000 were \$35,000, so her total sales were \$55,000 (\$20,000 + \$35,000).

210. **b.** Let x be the number of CDs that Matt has. Then, Tanya has $0.39x$ CDs, and Joan has $0.42(0.39x)$ CDs. This simplifies to $0.1638x$ CDs, so Joan has 16.38% as many CDs as Matt has.

211. **d.** First, find the sale price of the scarf and the gloves. They are both 20% off, which means that Christie paid 80% of the original price (100% – 20% = 80%). To find 80% of each price, multiply the price by the decimal equivalent of 80%: \$15.50 × 0.80 = \$12.40; \$5.50 × 0.80 = \$4.40. Together the two items cost \$16.80 (\$12.40 + \$4.40 = \$16.80). There is 5% sales tax on the total price. To find 5% of \$16.80, multiply \$16.80 by the decimal equivalent of 5%: \$16.80 × 0.05 = \$0.84. Therefore, Christie paid a total of \$17.64 (\$16.80 + \$0.84 = \$17.64).

212. **b.** To find 107% of \$54, multiply \$54 by the decimal equivalent of 107%: \$54 × 1.07 = \$57.78. If you chose **d**, you found what the cost of the book would be if it cost 7% *less* the next year.

213. **a.** If Larry earns a $3\frac{1}{4}$% (or 3.25%) raise, he will earn 103.25% of his original salary. To find 103.35% of \$32,000, multiply \$32,000 by the decimal equivalent of 103.25%: \$32,000 × 1.0325 = \$33,040. If you chose **d**, you found his salary with a 3% raise when multiplying by 1.03 or 0.03 and then adding that answer to his original salary.

214. a. Work backwards to find the answer. After lunch Bill had $8. He had spent 50% of what he had on lunch, and 50% is what is left. Since $8 is 50% of what he had before lunch, he had $16 before lunch. Using the same reasoning, $16 is 50% of what he had before buying school supplies. Therefore, he had $32 when he began shopping.

215. c. Let x be the original enrollment in yoga classes prior to January. Then, the enrollment at the end of January is $x + 0.037x = 1.037x$. Then, at the end of February, the enrollment is $1.037x - 0.0017(1.037x) = 1.019371x$. Finally, at the end of March, the enrollment is $1.019371x + 0.008(1.019371x) \approx 1.0275x$. So, the net change in enrollments is $1.0275x - x = 0.0275x$, which corresponds to a 2.75% net percent increase in enrollment during this three-month period.

216. d. Kristen has a total of 17% taken out of her check each week. Therefore, she is left with 83% of what she started with (100% − 17% = 83%). To find 83% of $550, multiply $550 by the decimal equivalent of 83%: $550 × 0.83 = $456.50.

217. d. Coastal Cable gained a total of 360,000 customers (1,800,000 − 1,440,000 = 360,000). To find out what percent of the original number of customers 360,000 represents, divide 360,000 by 1,440,000: 360,000 ÷ 1,440,000 = 0.25; 0.25 is equivalent to 25%. If you chose **c**, you found the percent of increase in relation to the new number of customers (1,800,000) rather than the original number of customers (1,440,000).

218. a. The price of heating oil rose $0.68 ($2.38 − $1.70 = $0.68). To find the percent of increase, divide $0.68 by the original cost of $1.70: $0.68 ÷ $1.70 = 0.4; 0.4 is equivalent to 40%. If you chose **c**, you found the percent of increase in relation to the new price ($2.38) rather than the original price ($1.70).

219. b. The percents must add to 100%; 24% + 13% + 41% = 78%. If 78% of the girls surveyed have been accounted for, the remainder of the girls must have said that field hockey is their favorite sport. To find the percent that said field hockey is their favorite sport, subtract 78% from 100%: 100% − 78% = 22%.

220. **c.** 78% of babies born weigh less than 8.5 pounds, but you must subtract the 25% that weigh less than 6 pounds: 78% – 25% = 53%. 53% of the babies born weigh between 6 and 8.5 pounds.

221. **d.** Find 20% of the original price of the coat and subtract it from the original price. To find 20%, multiply by 0.20: $80 × 0.20 = $16. Take $16 off the original price: $80 – $16 = $64. The first sale price is $64. Take 15% off this price using the same method: $64 × 0.15 = $9.60; $64 – $9.60 = $54.40. The new price of the coat is $54.40.

Another way of solving this problem is to look at the percent that is left after the discount has been taken. For example, if 20% is taken off, 80% is left (100% – 20%). Therefore, 80% of the original price is $80 × 0.80 = $64. If 15% is taken off this price, 85% is left: $64 × 0.85 = $54.40. This method eliminates the extra step of subtracting.

222. **b.** The percentage decrease corresponding to the first markdown is $\frac{250-210}{250} \times 100\% = 16\%$, while the percentage decrease corresponding to the second markdown is approximately $\frac{210-170}{210} \times 100\% \approx 19.05\%$. So, the percentage decrease for the second markdown is larger than the percentage decrease for the first.

223. **b.** Let x be the number of questions that Sue got correct. Set up a proportion of the form $\frac{100\%}{25 \text{ questions}} = \frac{76\%}{x \text{ questions}} : \frac{100\%}{25 \text{ questions}} = \frac{76\%}{x \text{ questions}}$. To solve for x, cross-multiply and then divide both sides by 100: $100x = 76(25) = 1{,}900$, so $x = 19$.

224. **a.** Write the relationship as a fraction: $\frac{part}{whole}$ or $\frac{walkers}{total} = \frac{8}{20}$. Find the decimal equivalent by dividing the numerator by the denominator (top ÷ bottom): $8 ÷ 20 = 0.4$. Change 0.4 to a percent by multiplying by 100: $0.4 × 100 = 40\%$.

Another way to look at this problem is using a proportion: $\frac{part}{whole} = \frac{\%}{100}$. You are looking for the percent, so that will be the variable: $\frac{8}{20} = \frac{x}{100}$. To solve the proportion, cross-multiply and set the answers equal to each other: $(8)(100) = 20x$. Solve for x by dividing both sides by 20.
$800 = 20x$
$\frac{800}{20} = \frac{20x}{20}$
$x = 40$
40% of the students in Kimmi's class walk to school.

225. **d.** First, find the number of walkers, and then find one third of that number. Find 37.5% of 24 by multiplying 24 by the decimal equivalent of 37.5%. To find the decimal equivalent, move the decimal point two places to the left; 37.5% = 0.375. Now, multiply 24 × 0.375 = 9. Find one third of 9 by dividing 9 by 3: 9 ÷ 3 = 3. Three walkers got rides to school today.

226. **c.** Find the price of the three items together (without tax): $35 + $45 + $20 = $100. Next, find 6% of $100. You can multiply $100 by 0.06, but it is easier to realize that 6% means "6 out of 100," so 6% of $100 is $6. The sales tax is $6.

A common mistake is to use 0.6 for 6% instead of 0.06; 0.6 is 60%. To find the decimal equivalent of a percent, you must move the decimal point two places to the left.

227. **a.** A proportion can be used to solve this problem: $\frac{part}{whole} = \frac{\%}{100}$. In this example, the *part* is the tax, the % is 8, and the *whole* is x. To solve the proportion, cross-multiply, set the cross-products equal to each other, and solve as shown below.

$\frac{2.12}{x} = \frac{8}{100}$

$(2.12)(100) = 8x$

$212 = 8x$

$\frac{212}{8} = \frac{8x}{8}$

$x = 26.5$

The price of the book was $26.50.

228. **c.** Find 17% of the bill by multiplying $65 by the decimal equivalent of 17%: $65 × 0.17 = $11.05. The tip should be $11.05.

229. **a.** Find 54% of 23,500 by multiplying 23,500 by the decimal equivalent of 54%: 23,500 × 0.54 = 12,690. 12,690 people are expected to vote for Mr. Salva.

230. **c.** The original price of the bike is 100%. If the sale takes 30% off the price, it will leave 70% of the original price (100% – 30% = 70%).

231. a. Let x be the number of adults attending the family reunion. "Forty percent of x" equals $0.40x$, and we are told that this is equal to 10. So, we have the equation $0.40x = 10$. To solve for x, divide both sides by 0.40: $\frac{10}{0.40} = 25$ adults.

232. c. Let x be the original price of a snow shovel. 250% of x equals $2.50x$, and we are told that this is equal to \$40. So, we have the equation $2.50x = \$40$. To solve for x, divide both sides by 2.50: $x = \frac{\$40}{2.50} = \$16$.

233. c. Find 32% of \$5,000 by multiplying \$5,000 by the decimal equivalent of 32%: $\$5,000 \times 0.32 = \$1,600$.

234. d. To determine this percentage, divide the amount of brown sugar Melinda has by the amount of brown sugar needed to make one entire recipe, and then multiply by 100%: $\frac{\frac{3}{4}}{2\frac{3}{8}} \times 100\% = (\frac{3}{4} \div \frac{19}{8}) \times 100\% = (\frac{3}{4} \times \frac{8}{19}) \times 100\% = \frac{6}{19} \times 100\% = 31.57\% \approx 32\%$.

235. b. Find 15% of 60 inches and add it to 60 inches. Find 15% by multiplying 60 by the decimal equivalent of 15%: $60 \times 0.15 = 9$. Add 9 inches to 60 inches to get 69 inches.

236. c. Call the original price of the jeans x. First 20% is deducted from the original cost (the original cost is 100%). 80% of the original cost is left (100% – 20% = 80%), and 80% of x is $0.80x$. The cost of the jeans after the first discount is $0.80x$. This price is then discounted 15%. Remember 15% is taken off the discounted price; 85% of the discounted price is left. Multiply the discounted price by 0.85 to find the price of the jeans after the second discount; $(0.85)(0.80x)$ is the cost of the jeans after both discounts. We are told that this price is \$80. Set the two expressions for the cost of the jeans equal to each other $(0.85)(0.80x) = \$80$ and solve for x (the original cost of the jeans).
$(0.85)(0.80x) = 80$
$0.68x = 80$
$\frac{0.68x}{0.68} = \frac{80}{0.68}$
$x = \$117.65$
The original price of the jeans was \$117.65.

237. **c.** Let x be the value of the baseball card last year. An increase in value of 325% is equal to $3.25x$, and we are told that this value is equal to $520. So, we have the equation $3.25x = \$520$. To solve for x, divide both sides by 3.25: $x = \frac{\$520}{3.25} = \$160$.

238. **c.** Break the rectangle into eighths as shown below. The shaded part is $\frac{6}{8}$ or $\frac{3}{4}$; $\frac{3}{4}$ is 75%.

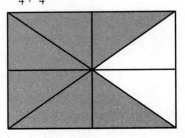

239. **a.** To find 20%, add 5% to 15%. Since 15% is known to be $42, 5% can be found by dividing $42 by 3 (15% ÷ 3 = 5%): $42 ÷ 3 = $14. To find 20%, add the 5% ($14) to the 15% ($42): $14 + $42 = $56. So, a 20% is tip $56.

240. **d.** Use a proportion to solve the problem: $\frac{part}{whole} = \frac{\%}{100}$. The *part* is $100,000, the *whole* is $130,000, and the *percentage* is x because it is unknown: $\frac{100,000}{130,000} = \frac{x}{100}$. To solve the proportion, cross-multiply, set the cross-products equal to each other, and solve as shown below.

$$\frac{100,000}{130,000} = \frac{x}{100}$$

$$(100)(100,000) = 130,000x$$

$$10,000,000 = 130,000x$$

$$\frac{10,000,000}{130,000} \times \frac{130,000x}{130,000}$$

$$x = 77$$

77% of the budget has been spent.

241. **b.** Multiply $359,000 by the decimal equivalent of 1.5% to find her commission: $359,000 × 0.015 = $5,385.

A common mistake is to use 0.15 for the decimal equivalent of 1.5%; 0.15 is equivalent to 15%. Remember, to find the decimal equivalent of a percent, move the decimal point two places to the left.

242. d. To find the price he sells it for, add the mark-up to his cost ($42). The mark-up is 110%. To find 110% of his cost, multiply by the decimal equivalent of 110%: $42 × 1.10 = $46.20. Add this mark-up to his cost to find the price the vase sells for: $46.20 + $42.00 = $88.20.

243. a. Let x be Bob's salary for the year. First, note that $1\frac{3}{4}\% = 1.75\%$. Now, a bonus of $1\frac{3}{4}\%$ of the salary for the year equals $0.0175x$, and we are told that this value is equal to $800. So, we have the equation $0.0175x = \$800$. To solve for x, divide both sides by 0.0175: $x = \frac{\$800}{0.0175} \approx \$45,714$.

244. d. Find the Social Security tax and the State Disability Insurance, and then subtract those amounts from Kyra's weekly wages. To find 7.51% of $895, multiply by the decimal equivalent of 7.51%: $895 × 0.0751 = $67.21 (rounded to the nearest cent). Next, find 1.2% of her wages by multiplying by the decimal equivalent of 1.2%: $895 × 0.012 = $10.74. Subtract $67.21 and $10.74 from Kyra's weekly wages of $895 to find the amount of her weekly paycheck: $895 − $67.21 − $10.74 = $817.05.

245. a. Find 5% of the bill by multiplying by the decimal equivalent of 5%: $178 × 0.05 = $8.90. They will save $8.90.
   A common mistake is to use 0.5 instead of 0.05 for 5%; 0.5 is 50%.

246. d. Find 30% of 1,800 by multiplying by the decimal equivalent of 30%: 1,800 × 0.30 = 540. The maximum number of calories from fats that Josephine can consume per day is 540.

247. c. Find 24% of $1,345 by multiplying by the decimal equivalent of 24%: $1,345 × 0.24 = $322.80. $322.80 can be deducted.

248. b. To compute 95% of a quantity, multiply that quantity by 0.95. Since Mark drinks 5 cups of coffee on Day 1, and then drinks 95% of the previous day's amount of coffee for each of the next 6 days (to reach Day 7), we must multiply the initial amount of coffee (5 cups) by 0.95 *six* times. This is represented by $(0.95)^6 × (5 \text{ cups}) \approx 3.7$ cups.

249. b. Multiply by the decimal equivalent of $\frac{1}{2}$% to find the amount of increase: $152,850 \times 0.005 = \$764.25$. This is how much sales increased. To find the actual amount of sales, add this increase to last month's total: $152,850 + \$764.25 = \$153,614.25$.

A common mistake is to use 0.5 (50%) or 0.05 (5%) for $\frac{1}{2}$%. Re-write $\frac{1}{2}$% as 0.5%. To find the decimal equivalent, move the decimal point two places to the left. This yields 0.005.

250. c. Find 5% of 220 by multiplying 220 by the decimal equivalent of 5%: $220 \times 0.05 = 11$ people.

A common mistake is to use 0.5 for 5%; 0.5 is actually 50%.

Algebra

Basic algebra problems require you to solve equations in which one or more elements are unknown. The unknown quantities are represented by variables, which are represented as letters of the alphabet, such as x or y. The questions in this chapter will give you practice in writing algebraic equations and using them to solve problems.

251. Assume that the number of hours Katie spent practicing soccer is represented by x. Michael practiced 4 hours less than twice the number of hours that Katie practiced. How many hours did Michael practice?
   a. $2x - 4$
   b. $2x + 4$
   c. $2x + 8$
   d. $4x + 4$

252. Patrick gets paid three dollars less than four times what Kevin gets paid. If the number of dollars that Kevin gets paid is represented by x, how many dollars does Patrick get paid?
   a. $3 - 4x$
   b. $3x - 4$
   c. $4x - 3$
   d. $4 - 3x$

253. If the expression $9y - 5$ represents a certain number, which of the following could NOT be the translation?

    a. five less than nine times y

    b. five less than the sum of 9 and y

    c. the difference between $9y$ and 5

    d. the product of nine and y, decreased by 5

254. During an expedition in the Appalachian Mountains, a group of adventurers traveled one-fourth of the way by foot, one-third of the way by canoe, and 8 miles by caravan. How many miles was the entire trip?

    a. 8.5 miles

    b. 15 miles

    c. 19.2 miles

    d. 21.4 miles

255. Frederick bought 11 books that cost d dollars each. What is the total cost of the books?

    a. $d + 11$

    b. $d - 11$

    c. $11d$

    d. $\frac{d}{11}$

256. There are m months in a year, w weeks in a month and d days in a week. How many days are there in a year?

    a. mwd

    b. $m + w + d$

    c. $\frac{mw}{d}$

    d. $d + \frac{w}{d}$

257. Carlie received x dollars for each hour she spent babysitting. She babysat a total of h hours. She then gave half of the money to a friend who had stopped by to help her. How much money did Carlie have after she had paid her friend?

    a. $\frac{hx}{2}$

    b. $\frac{x}{2} + h$

    c. $\frac{h}{2} + x$

    d. $2hx$

258. A long distance call costs x cents for the first minute and y cents for each additional minute. How much would a 10-minute call cost?
   a. $10xy$
   b. $x + 10y$
   c. $\frac{xy}{10}$
   d. $x + 9y$

259. Melissa is four times as old as Jim. Pat is 5 years older than Melissa. If Jim is y years old, how old is Pat?
   a. $4y + 5$
   b. $5y + 4$
   c. $4 \times 5y$
   d. $y + 5$

260. Sally is paid x dollars per hour for a 40-hour work week, and y dollars for each hour she works over 40 hours. How much did Sally earn last week if she worked 48 hours?
   a. $48xy$
   b. $40y + 8x$
   c. $40x + 8y$
   d. $48x + 48y$

261. Eduardo is combining two 6-inch pieces of wood with a piece that measures 4 inches to make one long board. How many total inches of wood does he have?
   a. 10 inches
   b. 16 inches
   c. 8 inches
   d. 12 inches

262. It takes 6 hours to fly against the wind from Baltimore to Los Angeles, and 5.2 hours to make the return trip with the wind. If an airplane averages 525 miles per hour in still air, what is the approximate speed of the wind? Round your answer to the nearest tenth.
   a. 34.5 miles per hour
   b. 37.5 miles per hour
   c. 40.5 miles per hour
   d. 42.5 miles per hour

263. Ten is decreased by four times the quantity of eight minus three. One is then added to that result. What is the final answer?
  a. ⁻5
  b. ⁻9
  c. 31
  d. ⁻8

264. The area of a square whose sides measure four units is added to the difference of eleven and nine divided by two. What is the total value?
  a. 9
  b. 16
  c. 5
  d. 17

265. An 1,800 calorie per day diet suggests eating a substantive breakfast, a light lunch, a reasonable dinner, and one snack. The calorie distribution is as follows: Breakfast has twice the number of calories as dinner, lunch has 80 fewer calories than dinner, and the snack is 100 calories. How many calories is breakfast?
  a. 365 calories
  b. 445 calories
  c. 890 calories
  d. 990 calories

266. John and Charlie have a total of 80 dollars. John has x dollars. How much money does Charlie have?
  a. $80
  b. $80 + x
  c. $80 − x
  d. x − $80

267. The temperature in Hillsville was 20° Celsius. What is the equivalent of this temperature in degrees Fahrenheit?
  a. 4 °F
  b. 43.1 °F
  c. 68 °F
  d. 132 °F

268. How many gallons of pure sulfuric acid must be mixed with three gallons of a solution that is 25% sulfuric acid to make a solution that is 60% sulfuric acid?

    a. 2.625 gallons

    b. 3.625 gallons

    c. 4.625 gallons

    d. 5.625 gallons

269. Celine deposited $505 into her savings account. If the interest rate of the account is 5% per year, how much interest will she have earned after 4 years?

    a. $252.50

    b. $606

    c. $10,100

    d. $101

270. A certain bank pays 3.4% interest per year for a certificate of deposit. What is the total balance of an account after 18 months with an initial deposit of $1,250?

    a. $765

    b. $2,015

    c. $63.75

    d. $1,313.75

271. Joe took out a car loan for $12,000. He paid $4,800 in interest at a rate of 8% per year. How many years did it take him to pay off the loan?

    a. 5

    b. 2.5

    c. 8

    d. 4

272. What is the annual interest rate on an account that earns $711 in simple interest over 36 months with an initial deposit of $7,900?

    a. 4.3%

    b. 3%

    c. 30%

    d. 4%

273. The speed of sound is approximately 760 miles per hour in still air. If a gas tank explodes $3\frac{3}{5}$ miles from your house, how long will it take the sound to reach you? Round your answer to the nearest second.
  a. Less than one second
  b. 17 seconds
  c. 80 seconds
  d. 211 seconds

274. A speed boat can maintain a constant speed of 14 miles per hour relative to the water. The boat makes a trip downstream in 25 minutes and the return trip upstream in 35 minutes. What is the speed of the water's current?
  a. $2\frac{1}{3}$ miles per hour
  b. $3\frac{1}{4}$ miles per hour
  c. $3\frac{3}{4}$ miles per hour
  d. 5 miles per hour

275. Jake needed to find the perimeter of an equilateral triangle whose sides measure $x + 4$ cm each. Jake realized that he could multiply $3(x + 4) = 3x + 12$ to find the total perimeter in terms of x. Which property did he use to multiply?
  a. Associative Property of Addition
  b. Distributive Property of Multiplication over Addition
  c. Commutative Property of Multiplication
  d. Inverse Property of Addition

276. The product of $^-5$ and a number is 35. What is the number?
  a. 40
  b. 30
  c. $^-7$
  d. 7

277. When ten is subtracted from the opposite of a number, the difference between them is five. What is the number?
  a. 15
  b. $^-15$
  c. $^-5$
  d. 5

278. The sum of ‾4 and a number is equal to ‾48. What is the number?
   a. ‾12
   b. ‾44
   c. 12
   d. ‾52

279. Twice a number increased by 11 is equal to 32 less than three times the number. What is the number?
   a. ‾21
   b. $\frac{21}{5}$
   c. 43
   d. $\frac{43}{5}$

280. If one is added to the difference when $10x$ is subtracted from ‾$18x$, the result is 57. What is the value of x?
   a. ‾2
   b. ‾7
   c. 2
   d. 7

281. If 0.3 is added to 0.2 times the quantity $x - 3$, the result is 2.5. What is the value of x?
   a. 1.7
   b. 26
   c. 14
   d. 17

282. Ted can power-wash both sides of the fence enclosing his yard in four hours. If he and his wife, Anne, work together, they can complete the same job in two hours. If Anne does the job alone, how would it take her to power-wash the fence?
   a. 3 hours
   b. 4 hours
   c. 5 hours
   d. 6 hours

283. The difference between six times the quantity $6x + 1$ and three times the quantity $x - 1$ is 108. What is the value of x?

   a. $\frac{12}{11}$

   b. $\frac{35}{11}$

   c. 12

   d. 3

284. Negative four is multiplied by the quantity $x + 8$. If $6x$ is then added to this, the result is $2x + 32$. What is the value of x?

   a. No solution

   b. Identity

   d. 0

   d. 16

285. A healthy weight range for a man is as follows:

     130 pounds for a height of 5 feet

     3 to 6 pounds per every inch of height beyond 5 feet

Which of the following represents a healthy weight range for a man whose height is 5 feet 11 inches?

   a. 97 pounds to 163 pounds

   b. 130 pounds to 196 pounds

   c. 97 pounds to 130 pounds

   d. 163 pounds to 196 pounds

286. Michael has 16 CDs. This is four more than one half the number of CDs that Kathleen has. How many CDs does Kathleen have?

   a. 6

   b. 24

   d. 4

   d. 12

287. The perimeter of a square can be expressed as $x + 4$. If one side of the square is 24, what is the value of x?

   a. 2

   b. 7

   c. 5

   d. 92

288. The perimeter of a rectangle is 21 inches. What is the measure of its width if its length is 3 inches greater than its width?
  a. 9 inches
  b. 3.75 inches
  c. 4.5 inches
  d. 3 inches

289. Bart's first four exam scores are 80, 76, 89, and 69. If the fifth exam counts twice, what must he score on the fifth exam to ensure that his exam average is between 80 and 85?
  a. Between 86 and 100
  b. Between 78 and 94
  c. Between 83 and 98
  d. Between 83 and 100

290. The sum of two consecutive even integers is 126. What are the integers?
  a. 62, 64
  b. 62, 63
  b. 60, 66
  d. 64, 66

291. Typical mark-ups on laptop computers during the holidays range from 10% to 20%. If the sticker price of a laptop is $1,300 prior to the holidays, which of the following is a range of the price during the holidays?
  a. $1,040 to $1,170
  b. $1,300 to $1,430
  c. $1,170 to $1,560
  d. $1,430 to $1,560

292. The sum of three consecutive even integers is 114. What is the value of the largest consecutive integer?
  a. 36
  b. 34
  c. 40
  d. 38

293. Two commuters leave the same city at the same time but travel in opposite directions. One car is traveling at an average speed of 63 miles per hour, and the other car is traveling at an average speed of 59 miles per hour. How many hours will it take before the cars are 610 miles apart?

    a. 4

    b. 6

    c. 30

    d. 5

294. Two trains leave the same city at the same time, one going east and the other going west. If one train is traveling at 65 mph and the other at 72 mph, how many hours will it take for them to be 822 miles apart?

    a. 9

    b. 7

    c. 8

    d. 6

295. Two trains leave two different cities 1,029 miles apart and head directly toward each other on parallel tracks. If one train is traveling at 45 miles per hour and the other at 53 miles per hour, how many hours will it take before the trains pass?

    a. 9.5

    b. 11

    c. 11.5

    d. 10.5

296. Nine minus five times a number, x, is no less than 39. Which of the following expressions represents all the possible values of the number?

    a. $x \leq 6$

    b. $x \geq {}^-6$

    c. $x \leq {}^-6$

    d. $x \geq 6$

297. Will has a bag of gumdrops. If he eats 2 of his gumdrops, there will be between 2 and 6 gumdrops left. Which of the following represents how many gumdrops, x, were originally in his bag?

a. $4 < x < 8$

b. $0 < x < 4$

c. $0 > x > 4$

d. $4 > x > 8$

298. The value of y is between negative three and positive eight, both inclusive. Which of the following represents y?

a. $-3 \leq y \leq 8$

b. $-3 < y \leq 8$

c. $-3 \leq y < 8$

d. $-3 \geq y \geq 8$

299. Five more than the quotient of a number and 2 is at least that number. What is the greatest value of the number?

a. 7

b. 10

c. 5

d. 2

300. Carl worked three more than twice as many hours as Cindy did. What is the maximum number of hours Cindy worked if together they worked 48 hours at most?

a. 17

b. 33

c. 37

d. 15

301. The cost of renting a bike at the local bike shop can be represented by the equation $y = 2x + 2$, where y is the total cost and x is the number of hours the bike is rented. Which of the following ordered pairs (x,y) would be a possible number of hours rented, x, and the corresponding total cost, y?

a. $(0, -2)$

b. $(2, 6)$

c. $(6, 2)$

d. $(-2, -6)$

302. A telephone company charges $0.35 for the first minute of a phone call and $0.15 for each additional minute of the call. Which of the following represents the cost, y, of a phone call lasting x minutes?

    a. $y = \$0.15(x - 1) + \$0.35$
    b. $x = \$0.15(y - 1) + \$0.35$
    c. $y = \$0.15x + \$0.35$
    d. $x = \$0.15y + \$0.35$

303. A ride in a taxicab costs $1.25 for the first mile and $1.15 for each additional mile. Which of the following could be used to calculate the total cost, y, of a ride that was x miles?

    a. $x = \$1.25(y - 1) + \$1.15$
    b. $x = \$1.15(y - 1) + \$1.25$
    c. $y = \$1.25(x - 1) + \$1.15$
    d. $y = \$1.15(x - 1) + \$1.25$

304. The cost of shipping a package through Shipping Express is $4.85 plus $2 per ounce of the weight of the package. Sally only has $10 to spend on shipping costs. Which of the following could Sally use to find the maximum number of ounces she can ship for $10?

    a. $\$4.85x + 2 \le \$10$
    b. $\$4.85x + 2 \ge \$10$
    c. $\$2x + \$4.85 \le \$10$
    d. $\$2x + \$4.85 \ge \$10$

305. Green Bank charges a monthly fee of $3 for a checking account and $0.10 per check. Savings-R-Us bank charges a $4.50 monthly fee and $0.05 per check. How many checks need to be used for the monthly costs to be the same for both banks?

    a. 25
    b. 30
    c. 35
    d. 100

306. Easy Rider taxi service charges a pick-up fee of $2 and $1.25 for each mile. Luxury Limo taxi service charges a pick-up fee of $3.25 and $1 per mile. How many miles need to be driven for both services to cost the same amount?
   a. 24
   b. 12
   c. 10
   d. 5

307. The sum of two integers is 36, and their difference is 6. What is the smaller of the two numbers?
   a. 21
   b. 15
   c. 16
   d. 18

308. One integer is two more than another. The sum of the lesser integer and twice the greater is 7. What is the greater integer?
   a. 1
   b. 2
   c. 3
   d. 7

309. During a "buyer's market," a house will typically sell for 86% to 95% of its estimated worth. If the estimated value of a house is P dollars, which of the following reflects the range of the amount of money lost (by the seller) in selling the house in a "buyer's market"?
   a. $0.05P$ to $0.14P$
   b. $5P$ to $14P$
   c. $86P$ to $95P$
   d. $0.86P$ to $0.95P$

310. A house's thermostat is known to have a margin of error of ±3%. If the temperature reading is 72 °F, what is the range of possible temperatures in the house?
   a. 72 °F to 74.16 °F
   b. 69.84 °F to 74.16 °F
   c. 69.84 °F to 72 °F
   d. 69 °F to 75 °F

311. The perimeter of a rectangle is 104 inches. The width is 6 inches less than 3 times the length. Find the width of the rectangle.
    a. 13.5 inches
    b. 37.5 inches
    c. 14.5 inches
    d. 15 inches

312. If a person's blood alcohol level (call it A) registers at least 0.07, he or she will be arrested for driving under the influence. If the breathalyzer test accurately measures blood alcohol level within ±0.1%, which of the following inequalities provides a range of actual blood alcohol levels that will not result in an arrest?
    a. $A < 0.071$
    b. $A < 0.17$
    c. $A < 0.07$
    d. $A < 0.069$

313. Jackie invested money in two different accounts, one of which earned 12% interest per year and another that earned 15% interest per year. The amount invested at 15% was $100 more than twice the amount at 12%. How much was invested at 12% if the total annual interest earned was $855?
    a. $4,100
    b. $2,100
    c. $2,000
    d. $4,000

314. Kevin invested $4,000 in an account that earns 6% interest per year and x dollars in a different account that earns 8% interest per year. How much is invested at 8% if the total amount of interest earned annually is $405.50?
    a. $2,075.00
    b. $4,000.00
    c. $2,068.75
    d. $2,075.68

315. Megan bought x pounds of coffee that cost $3 per pound, and 18 pounds of coffee at $2.50 per pound for the company picnic. Find the total number of pounds of coffee she purchased if the average cost per pound of both types together is $2.85.
   a. 42
   b. 18
   c. 63
   d. 60

316. The student council bought two different types of candy for the school fair. They purchased 40 pounds of candy at $2.15 per pound and x pounds at $1.90 per pound. What is the total number of pounds they bought if the total amount of money spent on candy was $162?
   a. 42
   b. 40
   c. 80
   d. 52

317. The manager of a garden store ordered two different kinds of marigold seeds for her display. The first type cost $1 per packet, and the second type cost $1.26 per packet. How many packets of the first type did she purchase if she bought 50 more of the $1.26 packets than the $1 packets and spent a total of $402?
   a. 150
   b. 200
   c. 250
   d. 100

318. Harold used a 3% iodine solution and a 20% iodine solution to make a 85-ounce solution that was 19% iodine. How many ounces of the 3% iodine solution did he use?
   a. 5
   b. 80
   c. 60
   d. 20

319. A chemist mixed a solution that was 34% acid with another solution that was 18% acid to produce a 30-ounce solution that was 28% acid. How many ounces of the 34% acid solution did he use?

a. 27

b. 11.25

c. 18.75

d. 28

320. Bob is 2 years away from being twice as old as Ellen. The sum of twice Bob's age and three times Ellen's age is 66. How old is Ellen?

a. 15

b. 10

c. 18

d. 20

321. Sam's age is 1 year less than twice Shari's age. The sum of their ages is 92. How old is Sam?

a. 48

b. 31

c. 65

d. 61

322. Ted spent $\frac{3}{8}$ of his Saturday leisure time watching reruns on TV, $\frac{3}{16}$ of his time cleaning, $\frac{1}{4}$ of his time reading a novel, and 30 minutes surfing the Internet. How many minutes did he spend watching TV?

a. 15 minutes

b. 30 minutes

c. 60 minutes

d. 105 minutes

323. Sue is three times as old as Karen, Karen is one year older than Mary, and Mary's age is five years less than twice Danielle's age. Which of the following expressions represents Sue's age in terms of Danielle's age?

a. $6d - 12$

b. $6d - 15$

c. $2d - 1$

d. $15d - 3$

324. The price of a student ticket is $1 more than half the price of an adult ticket. Six adult tickets and four student tickets cost $76. What is the price of one adult ticket?
    a. $2.50
    b. $9.00
    c. $5.50
    d. $4.00

325. An Internet site has a sale on MP3 downloads. The first ten downloads each cost x dollars. The next sixteen downloads each cost $\frac{x}{2}$ dollars, and each one beyond these initial twenty-six downloads each cost $\frac{x}{4}$ dollars. Which expression below represents the cost for 32 MP3 downloads?
    a. $(26 + 1.5x)$ dollars
    b. $(32 + 1.75x)$ dollars
    c. $19.5x$ dollars
    d. $(26 + x)$ dollars

326. Noel rode $3x$ miles on his bike, and Jamie rode $5x$ miles on hers. In terms of x, what is the total number of miles they both rode?
    a. $15x$ miles
    b. $15x^2$ miles
    c. $8x$ miles
    d. $8x2$ miles

327. A piggy bank is full of only nickels and dimes. If the bank contains 65 coins with a total value of $5.00, determine how many dimes are in the bank.
    a. 30
    b. 35
    c. 40
    d. 45

328. Laura has a rectangular garden whose width is x^3 and whose length is x^4. In terms of x, what is the area of her garden?
    a. $2x^7$
    b. x^7
    c. x^{12}
    d. $2x^{12}$

329. The base b and height h of a collection of triangles are related by the formula $h = (\frac{1}{2}b + 2)^2$. Which of the following pairs of dimensions describe a triangle in this collection?

    a. $b = 1, h = 5$
    b. $b = 4, h = 4$
    c. $b = 3, h = 12.25$
    d. $b = 2, h = 6$

330. The area of a parallelogram is x^8. If the base is x^4, what is the height in terms of x?

    a. x^4
    b. x^2
    c. x^{12}
    d. x^{32}

331. An industrial strength snow blower is worth \$900. The value depreciates as time goes on so that its value V after m months is given by $V = \$900 - \$20m$, where $0 \le m \le 45$. After how many months will the snow blower be worth \$300?

    a. 5
    b. 15
    c. 30
    d. 45

332. The product of $6x^2$ and $4xy^2$ is divided by $3x^3y$. What is the simplified expression?

    a. $8y$
    b. $\frac{4y}{x}$
    c. $4y$
    d. $\frac{8y}{x}$

333. If the side of a square can be expressed as a^2b^3, what is the area of the square in simplified form?

    a. a^4b^5
    b. a^4b^6
    c. a^2b^6
    d. a^2b^5

334. If $3x^2$ is multiplied by the quantity $2x^3y$ raised to the fourth power, what would this expression simplify to?

    a. $48x^{14}y^4$

    b. $1,296x^{16}y^4$

    c. $6x^9y^4$

    d. $6x^{14}y^4$

335. Sara's bedroom is in the shape of a rectangle. The dimensions are $2x$ and $4x + 5$. What is the area of Sara's bedroom?

    a. $18x$

    b. $18x^2$

    c. $8x^2 + 5x$

    d. $8x^2 + 10x$

336. You just bought a new front-loading washing machine for $1,200. The value depreciates as time goes on so that its value V after m months is given by $V = \$1,200 - \$20m$. What will be the value of the washing machine after 15 months?

    a. $1,000

    b. $950

    c. $925

    d. $900

337. A number, x, increased by 3 is multiplied by the same number, x, increased by 4. What is the product of the two numbers in terms of x?

    a. $x^2 + 7$

    b. $x^2 + 12$

    c. $x^2 + 7x + 12$

    d. $x^2 + x + 7$

338. The length of Kara's rectangular patio can be expressed as $2x - 1$, and the width can be expressed as $x + 6$. In terms of x, what is the area of her patio?

    a. $2x^2 + 13x - 6$

    b. $2x^2 - 6$

    c. $2x^2 - 5x - 6$

    d. $2x^2 + 11x - 6$

339. A car travels at a rate of $(4x^2 - 2)$ miles per hour. What is the distance in miles that this car will travel in $(3x - 8)$ hours?
  a. $12x^3 - 32x^2 - 6x + 16$
  b. $12x^2 - 32x^2 - 6x + 16$
  c. $12x^3 + 32x^2 - 6x - 16$
  d. $12x^3 - 32x^2 - 5x + 16$

340. The area of the base of a prism can be expressed as $x^2 + 4x + 1$, and the height of the prism can be expressed as $x - 3$. What is the volume of this prism in terms of x?
  a. $x^3 + x^2 - 13x - 3$
  b. $x^3 + 7x^2 - 13x - 3$
  c. $x^3 - x^2 - 11x - 3$
  d. $x^3 + x^2 - 11x - 3$

341. The dimensions of a rectangular prism can be expressed as $x + 1$, $x - 2$, and $x + 4$. In terms of x, what is the volume of the prism?
  a. $x^3 + 3x^2 + 6x - 8$
  b. $x^3 + 3x^2 - 6x - 8$
  c. $x^3 + 5x^2 - 2x + 8$
  d. $x^3 - 5x^2 - 2x - 8$

342. The area of Mr. Smith's rectangular classroom is $x^2 - 36$. Which of the following binomials could represent the length and the width of the room?
  a. $(x + 6)(x + 6)$
  b. $(x - 6)(x - 6)$
  c. $(x + 6)(x - 6)$
  d. $x(x - 36)$

343. The area of a parallelogram can be expressed as the binomial $2x^2 - 10x$. Which of the following could be the length of the base and the height of the parallelogram?
  a. $2x(x^2 - 5x)$
  b. $2x(x - 5)$
  c. $(2x - 1)(x - 10)$
  d. $(2x - 5)(x + 2)$

344. A farmer's rectangular field has an area that can be expressed as the trinomial $x^2 + 2x + 1$. In terms of x, what are the dimensions of the field?

    a. $(x + 1)(x + 2)$
    b. $(x - 1)(x - 2)$
    c. $(x - 1)(x + 2)$
    d. $(x + 1)(x + 1)$

345. A summer camp enrolls 540 children each summer. If the ratio of boys to girls this summer is 7 to 5, how many more boys than girls are attending the camp?

    a. 90
    b. 225
    c. 275
    d. 315

346. Jake has a pile of nickels, dimes, and quarters. He has three more dimes than nickels and twice as many quarters as dimes. Which of the following expressions represents the total value of Jake's collection of coins (in dollars)?

    a. $[x + (x + 3) = (2x + 6)]$
    b. $\$0.05x + \$0.10(x + 3) + \$0.25(2x + 6)$
    c. $\$5x + \$10(x + 3) + \$25(2x + 6)$
    d. $\$0.40\,[x + (x + 3) + (2x + 6)]$

347. There are 24 rolls of toilet paper in a mega-pack at the local grocery store. If each roll uses 15 square inches of cardboard, how many square inches of cardboard are used in y mega-packs?

    a. y
    b. $15y$
    c. $24y$
    d. $360y$

348. A magazine company makes it a point to include 1.1 times as many featured stories in the current issue as were included in the previous month's issue. If the present month's issue contains 15 featured stories, which of the following expressions can be used to determine approximately how many featured stories will be in the issue y months from now?

    a. $(1.1)^y \times 15$

    b. $(1.1 + y) \times 15$

    c. $(1.1\,y) \times 15$

    d. $(y^{1.1}) \times 15$

349. There are 5,000 tickets available on Day 1 for sale to a rock concert. The number of available tickets decreases by m tickets each day for the next two weeks, then by $2m$ each day after until there are none left. Which of the following expressions represents the number of tickets available on Day 21?

    a. $14m$

    b. $28m$

    c. $5,000 - 28m$

    d. $5,000 - 14m$

350. It takes light 5.3×10^{-6} seconds to travel one mile. What is this time in standard notation?

    a. 0.00000053

    b. 0.000053

    c. 5.300000

    d. 0.0000053

351. A local hardware store's profit P (measured in dollars) earned by selling indoor grills is given by the equation $P = 5g^2 - 1,125$, where g is the number of grills sold. How many grills must be sold to break even (that is, earn zero dollars in profit)?

    a. 5

    b. 15

    c. 50

    d. 225

352. The square of a number added to 25 equals 10 times the number. What is the number?

a. $^-5$

b. 10

c. $^-10$

d. 5

353. The sum of the square of a number and 12 times the number is $^-27$. What is the smaller of the two numbers that satisfy this description?

a. $^-3$

b. $^-9$

c. 3

d. 9

354. Scott listens to MP3s for x hours per day. The average duration of a single MP3 file is 3 minutes 30 seconds. Which of the following expressions represents the average number of MP3s he listens to during this amount of time?

a. $\frac{x}{3.5}$

b. $\frac{3.5}{x}$

c. $\frac{60x}{3.5}$

d. $\frac{3.5}{60x}$

355. A Girl Scout has a stack of $1, $5, and $10 bills that she obtained while collecting money for Girl Scout cookies. She has two fewer $10 bills than $5 bills, and four times as many $1 bills as $5 bills. If she has a total of $170, how many $1 bills does she have?

a. 20

b. 22

c. 40

d. 88

356. The number of 5-gallon barrels of paint needed to cover the exterior of a house with one coat of paint is given by the expression $[8s(w + l) - 6s] \div 397$, where s is the number of stories, l is the length of the house, and w is the width of the house. How many gallons of paint are needed to apply two coats to a 35 foot × 65 foot, two-story house?

    a. 4

    b. 8

    c. 20

    d. 40

357. A worker can stuff E envelopes per minute. A second, more expedient worker can stuff $3E$ envelopes per minute. How many minutes will it take the two workers, working together, to stuff 5,000 envelopes?

    a. $\dfrac{1,250}{E}$

    b. $\dfrac{E}{20,000}$

    c. $\dfrac{E}{1,250}$

    d. $20,000E$

358. A 60-foot piece of rope is cut into 3 pieces. The second piece must be 1 foot shorter than twice the length of the first piece, and the third piece must be 10 feet longer than three times the length of the second piece. How long should the longest piece be?

    a. 6 feet

    b. 11 feet

    c. 43 feet

    d. 60 feet

359. At Zides Sport Shop, a canister of Wilson tennis balls costs $3.50, and a canister of Penn tennis balls costs $2.75. The high school tennis coach bought canisters of both brands of balls, spending exactly $40.25 before the sales tax. If he bought one more canister of Penn balls than he did Wilson balls, how many canisters of each did he purchase?

    a. 5 canisters of Wilson balls and 6 canisters of Penn balls

    b. 6 canisters of Wilson balls and 7 canisters of Penn balls

    c. 7 canisters of Wilson balls and 8 canisters of Penn balls

    d. 8 canisters of Wilson balls and 9 canisters of Penn balls

360. One essential procedure needed to ensure the success of a Microgravity Germination Project is that 10 gallons of a 70% concentrated nitrogen solution be administered to the bean seeds. If the Payload Specialist has some 90% nitrogen and some 30% nitrogen, how many gallons of 30% nitrogen should she use when combining them in order to obtain the desired solution?

    a. $3\frac{1}{3}$

    b. 5

    c. $6\frac{2}{3}$

    d. $7\frac{3}{4}$

361. What is the lesser of two consecutive positive integers whose product is 90?

    a. ⁻9

    b. 9

    c. ⁻10

    d. 10

362. Two brothers are reading Carl Sagan's *Contact*, but one of them is 4 pages behind the other. If they add their current page numbers together, the sum is 408. On what page is the slower reader?

    a. 198

    b. 202

    c. 206

    d. 210

363. Two ships leave the same port precisely at noon. One travels due east, and the other travels due west. The ship traveling west travels at an average speed that is 10 miles per hour slower than the one traveling east. At 3:30 P.M., the two ships are 175 miles apart. How many miles has the eastbound ship traveled during this time?

    a. 105 miles

    b. 95 miles

    c. 80 miles

    d. 70 miles

364. Mike took three Biology exams and has an average score of 88. His second exam score was ten points better than his first, and his third exam was 4 points better than his second exam's score. What were his three exam scores?
   a. 65, 75, and 79
   b. 82, 92, and 96
   c. 80, 90, and 94
   d. 75, 85, and 89

365. The sum of the squares of two consecutive positive odd integers is 74. What is the value of the smaller integer?
   a. 3
   b. 7
   c. 5
   d. 11

366. If the difference between the squares of two consecutive integers is 15, find the value of the larger integer.
   a. 8
   b. 7
   c. 6
   d. 9

367. The square of one integer is 55 less than the square of the next consecutive integer. Find the value of the lesser integer.
   a. 23
   b. 24
   c. 27
   d. 28

368. A 4-inch by 6-inch photograph is going to be enlarged by increasing each side by the same amount. The new area of the photo is 168 square inches. By how many inches is each dimension increased?
   a. 12
   b. 10
   c. 8
   d. 6

369. A photographer decides to reduce a picture she took in order to fit it into a certain frame. She needs the new picture to be one-third of the area of the original. If the original picture was 4 inches by 6 inches, how many inches is the smaller dimension of the reduced picture if each dimension changes by the same amount?
   a. 2
   b. 3
   c. 4
   d. 5

370. A rectangular garden has a width of 20 feet and a length of 24 feet. If each side of the garden is increased by the same amount, how many feet is the new length if the new area is 141 square feet more than the original?
   a. 23
   b. 24
   c. 26
   d. 27

371. Ian can remodel a kitchen in 20 hours and Jack can do the same job in 15 hours. If they work together, approximately how many hours will it take them to remodel the kitchen?
   a. 5.6
   b. 8.6
   c. 7.5
   d. 12

372. Peter can paint a room in an hour and a half and Joe can paint the same room in 2 hours. How many minutes will it take them to paint the room if they do it together? Round your answer to nearest minute.
   a. 51
   b. 64
   c. 30
   d. 210

373. Carla can plant a garden in 3 hours and Charles can plant the same garden in 4.5 hours. If they work together, how many hours will it take them to plant the garden?
   a. 1.5
   b. 2.1
   c. 1.8
   d. 7.5

374. If Jim and Jerry work together they can finish a job in 4 hours. If it takes Jim 10 hours to finish the job when working alone, how many hours would it take Jerry to do the job alone? Round to the nearest tenth of an hour.
   a. 16
   b. 5.6
   c. 6.7
   d. 6.0

375. Bill and Ben can clean the garage together in 6 hours. If it takes Bill 15 hours working alone, how long will it take Ben working alone?
   a. 11 hours
   b. 9 hours
   c. 16 hours
   d. 10 hours

Answer Explanations

The following explanations show one way in which each problem can be solved. You may have another method for solving these problems.

251. **a.** The translation of "twice the number of hours" is $2x$. Four hours less than $2x$ becomes $2x - 4$.

252. **c.** When the key phrase *less than* appears in a sentence, it means that you will subtract from the next part of the sentence, so it will appear at the end of the expression. "Four times a number" is equal to $4x$ in this problem. Three less than $4x$ is $4x - 3$.

253. **b.** Each one of the answer choices would translate to $9y - 5$ except for choice **b.** The word *sum* is a key word for addition, and $9y$ is equal to "9 times y."

254. **c.** Let x be the total distance of the trip (in miles). We are given that the group traveled $\frac{1}{4}x$ miles by foot, $\frac{1}{3}x$ miles by canoe, and 8 miles by caravan. The sum of these three distances is x. This yields the equation $\frac{1}{4}x + \frac{1}{3}x + 8 = x$. To solve for x, first clear the fractions by multiplying both sides by the least common denominator of all fractions occurring in the equation, namely 12: $3x + 4x + 96 = 12x$. Next, gather like terms on the left-side: $7x + 96 = 12x$. Next, subtract $7x$ from both sides, and then divide both sides by 5: $96 = 5x$, so $x = \frac{96}{5} = 19.2$ miles.

255. **c.** Frederick would multiply the number of books, 11, by how much each one costs, d. For example, if each one of the books cost $10, he would multiply 11 times $10 and get $110. Therefore, the answer is $11d$.

256. **a.** In this problem, multiply d and w to get the total days in one month and then multiply that result by m to get the total days in the year. This can be expressed as mwd, which means m times w times d.

257. **a.** To calculate the total amount of money she received, multiply x dollars per hour times h, the number of hours she worked. This becomes xh. Divide this amount by 2 since she gave half to her friend. Thus, $\frac{xh}{2}$ is how much money she has left.

258. d. The cost of the call is x cents plus y times the additional minutes. Since the call is 10 minutes long, the caller will pay x cents for 1 minute and y cents for the other nine. Therefore the expression is $1x + 9y$, or $x + 9y$, since it is not necessary to write a 1 in front of a variable.

259. a. Start with Jim's age, y, since he is the youngest. Melissa is four times as old as he is, so her age is $4y$. Pat is 5 years older than Melissa, so Pat's age would be Melissa's age, $4y$, plus another 5 years. Thus, Pat's age is $4y + 5$.

260. c. Since she worked 48 hours, Sally will get paid her regular amount, x dollars, for 40 hours and a different amount, y, for the additional 8 hours. This becomes 40 *times* x plus 8 *times* y, which translates to $40x + 8y$.

261. b. This problem translates to the expression $6 \times 2 + 4$. Using order of operations, do the multiplication first: $6 \times 2 = 12$, then add $12 + 4 = 16$ inches.

262. b. Let x be the average speed of the wind. We will use the formula *distance = rate × time* to obtain expressions for the distance from Baltimore to Los Angeles, and the reverse trip.

<u>Baltimore to Los Angeles</u>: Since the plane is flying against the wind, the speed of the plane is $(525 - x)$ miles per hour. We are given that the time to make this trip is 6 hours, so an expression for the distance traveled is $(525 - x)(6)$ miles.

<u>Los Angeles to Baltimore</u>: Since the plane is flying with the wind, the speed of the plane is $(525 + x)$ miles per hour. We are given that the time to make the trip is 5.2 hours. So, an expression for the distance traveled is $(525 + x)(5.2)$ miles.

The distances traveled for both trips are the same, so that we obtain the equation $(525 - x)(6) = (525 + x)(5.2)$.

To solve for x, simplify the left and right sides: $3{,}150 - 6x = 2{,}730 + 5.2x$. Next, add $6x$ to both sides and subtract $2{,}730$ from both sides: $420 = 11.2x$. Finally, divide both sides by 11.2: $x = \frac{420}{11.2} = 37.5$ miles per hour.

263. b. This problem translates to the expression $10 - 4(8 - 3) + 1$. Using order of operations, do the operation inside the parentheses first: $10 - 4(5) + 1$. Since multiplication is next, multiply 4×5: $10 - 20 + 1$. Add and subtract in order from left to right: $10 - 20 = {}^-10$; ${}^-10 + 1 = {}^-9$.

264. d. This problem translates into the expression $4^2 + (11 - 9) \div 2$. Using order of operations, do the operation inside the parentheses first: $4^2 + 2 \div 2$. Evaluate the exponent: $16 + 2 \div 2$. Divide $2 \div 2$: $16 + 1$. Add: $16 + 1 = 17$.

265. c. Let x be the number of calories for dinner. Then, breakfast is $2x$ calories and lunch is $(x - 80)$ calories. Adding the numbers of calories for breakfast, lunch, dinner, and snack yields the following equation:

$$2x + (x - 80) + x + 100 = 1{,}800$$

To solve for x, simplify the left side: $4x + 20 = 1{,}800$. Next, subtract 20 from both sides: $4x = 1{,}780$. Finally, divide both sides by 4: $x = 445$. So, breakfast is $2(445) = 890$ calories

266. c. If the total amount of money for both is \$80, then the amount for one person is \$80 minus the amount of the other person. Since John has x dollars, Charlie's amount is $\$80 - x$.

267. c. Use the formula $F = \frac{9}{5}C + 32$. Substitute the Celsius temperature of 20° for C in the formula. This results in the equation $F = \frac{9}{5}(20) + 32$. Following the order of operations, multiply $\frac{9}{5}$ and 20 to get 36. The final step is to add $36 + 32$ for an answer of 68 °F.

268. a. Let x be the number of gallons of pure sulfuric acid needed. Multiply the concentration of sulfuric acid by the number of gallons for both solutions, and add them to obtain an equation involving x:

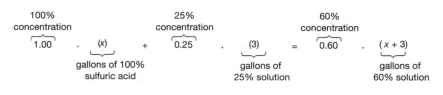

To solve for x, simplify the left and right-sides: $x + 0.75 = 0.60x + 1.80$. Next, subtract $0.60x$ and 0.75 from both sides: $0.40x = 1.05$. Finally, divide both sides by 0.40: $x = \frac{1.05}{0.40} = 2.625$ gallons.

269. d. Using the simple interest formula *Interest = principal × rate × time*, or $I = prt$, substitute $p = \$505$, $r = 0.05$ (the interest rate as a decimal), and $t = 4$: $I = (505)(0.05)(4)$. Multiply to get a result of $I = \$101$.

270. d. Using the simple interest formula *Interest = principal × rate × time*, or $I = prt$, substitute $p = \$1{,}250$, $r = 0.034$ (the interest rate as a decimal), and $t = 1.5$ (18 months is equal to 1.5 years): $I = (1{,}250)(0.034)(1.5)$. Multiply to get a result of $I = \$63.75$. To find the total amount in the account after 18 months, add the interest to the initial principal: $\$63.75 + \$1{,}250 = \$1{,}313.75$.

271. a. Using the simple interest formula *Interest = principal × rate × time*, or $I = prt$, substitute $I = \$4{,}800$, $p = \$12{,}000$, and $r = 0.08$ (the interest rate as a decimal): $\$4{,}800 = (\$12{,}000)(0.08)(t)$. Multiply $\$12{,}000$ and 0.08 to get $\$960$, so $\$4{,}800 = \$960t$. Divide both sides by $\$960$ to get $5 = t$. Therefore, it took Joe 5 years to pay off the loan.

272. b. Using the simple interest formula *Interest = principal × rate × time*, or $I = prt$, substitute $I = \$711$, $p = \$7{,}900$, and $t = 3$ (36 months is equal to 3 years): $711 = (\$7{,}900)(r)(3)$. Multiply $\$7{,}900$ and 3 on the right side to get a result of $\$711 = \$23{,}700r$. Divide both sides by $\$23{,}700$ to get $r = 0.03$, which is a decimal equal to 3%.

273. b. Divide the number of miles between the house and gas tank explosion ($3\frac{3}{5}$ miles) by the speed of sound (760 miles per hour) to determine the number of hours it takes the sound to reach the house: $\frac{35}{60} \approx 0.00474$. Now, convert to seconds using the fact that 1 minute = 60 seconds and 1 hour = 60 minutes:

0.00474 hours	60 minutes	60 seconds
	1 hour	1 minute

Multiply the numbers across the top and bottom, and cancel pairs of like units in the top and bottom to conclude that

$$0.00474 \text{ hours} \approx 17.064 \text{ seconds}.$$

So, it takes approximately 17 seconds for the sound of the explosion to reach the house.

274. **a.** Let x be the speed of the current (in miles per hour). We will use the formula *distance = rate × time* to obtain expressions for the distance traveled for the upstream and downstream trips.

<u>Upstream</u>: Since the speed boat is moving against the current, the speed of the boat is $(14 - x)$ miles per hour. We are given that the time to make this trip is 35 minutes, which equals $\frac{35}{60} = \frac{7}{12}$ hours. So, an expression for the distance traveled upstream is $(14 - x)(\frac{7}{12})$ miles.

<u>Downstream</u>: Since the speed boat is moving with the current, the speed of the boat is $(14 + x)$ miles per hour. We are given that the time to make this trip is 25 minutes, which equals $\frac{25}{60} + \frac{5}{12}$ hours. So, an expression for the distance traveled downstream is $(14 + x)(\frac{5}{12})$ miles.

The distances traveled for both trips are the same, so we obtain the equation $(14 - x)(\frac{7}{12}) = (14 + x)(\frac{5}{12})$.

To solve for x, first clear the fractions by multiplying both sides of the equation by 12: $(14 - x)(7) = (14 + x)(5)$. Next, simplify the left and right sides: $98 - 7x = 70 + 5x$. Next, subtract 70 and add $7x$ to both sides: $12x = 28$. Finally, divide both sides by 12: $x = \frac{28}{12} = 2\frac{4}{12} = 2\frac{1}{3}$ miles per hour. So, the speed of the current is $2\frac{1}{3}$ miles per hour.

275. **b.** In the statement, 3 is being multiplied by the quantity in the parentheses, $x + 4$. The distributive property allows you to multiply $3 \times x$ and add it to 3×4, simplifying to $3x + 12$.

276. **c.** Let y = the unknown number. The word *product* is a key word for multiplication. Therefore the equation is $^-5y = 35$. To solve this, divide each side of the equation by $^-5$: $\frac{^-5y}{^-5} = \frac{35}{^-5}$. The variable is now isolated: $y = ^-7$.

277. **b.** Let x = the unknown number. The opposite of this number is $-x$. The words *subtraction* and *difference* both tell you to subtract, so the equation becomes $-x - 10 = 5$. To solve this, add 10 to both sides of the equation: $-x - 10 + 10 = 5 + 10$. Simplify to $-x = 15$. Divide both sides of the equation by $^-1$. Remember that $-x = {}^-1x$; $\frac{-x}{^-1} = \frac{^-15}{^-1}$. The variable is now isolated: $x = {}^-15$.

278. **b.** Let x = the unknown number. Since *sum* is a key word for addition, the equation is $^-4 + x = {}^-48$. Add 4 to both sides of the equation; $^-4 + 4 + x = {}^-48 + 4$. The variable is now isolated: $x = {}^-44$.

279. **c.** Let x = the unknown number. Now translate each part of the sentence. Twice a number increased by 11 is $2x + 11$; 32 less than 3 times a number is $3x - 32$. Set the expressions equal to each other: $2x + 11 = 3x - 32$. Subtract $2x$ from both sides of the equation: $2x - 2x + 11 = 3x - 2x - 32$. Simplify: $11 = x - 32$. Add 32 to both sides of the equation: $11 + 32 = x - 32 + 32$. The variable is now isolated: $x = 43$.

280. **a.** The statement, "If one is added to the difference when $10x$ is subtracted from ^-18x, the result is 57," translates to the equation $^-18x - 10x + 1 = 57$. Combine like terms on the left side of the equation: $^-28x + 1 = 57$. Subtract 1 from both sides of the equation: $^-28x + 1 {}^-1 = 57 - 1$. Divide each side of the equation by $^-28$: $\frac{-28x}{-28} = \frac{56}{-28}$. The variable is now isolated: $x = {}^-2$.

281. **c.** The statement, "If 0.3 is added to 0.2 times the quantity $x - 3$, the result is 2.5," translates to the equation $0.2(x - 3) + 0.3 = 2.5$. Remember to use parentheses for the expression when the words *the quantity* are used. Use the distributive property on the left side of the equation: $0.2x - 0.6 + 0.3 = 2.5$. Combine like terms on the left side of the equation: $0.2x + {}^-0.3 = 2.5$. Add 0.3 to both sides of the equation: $0.2x + {}^-0.3 + 0.3 = 2.5 + 0.3$. Simplify: $0.2x = 2.8$. Divide both sides by 0.2: $\frac{0.2x}{0.2} = \frac{2.8}{0.2}$. The variable is now isolated: $x = 14$.

282. b. Let x be the number of hours it takes Anne to complete the job on her own.

Since Ted can do the job in 4 hours, he can do $\frac{1}{4}$ of the job in 1 hour. Similarly, Anne can complete $\frac{1}{x}$ of the job in 1 hour. As such, working together, they complete $\frac{1}{4} + \frac{1}{x}$ of the job in 1 hour. We are given that Ted and Anne can finish the job in 2 hours when they work together, so that they complete $\frac{1}{2}$ of the job in 1 hour. We now have two expressions for the portion of the job they complete, working together, in 1 hour; equate them to obtain an equation in terms of x: $\frac{1}{4} + \frac{1}{x} = \frac{1}{2}$.

To solve for x, clear the fractions by multiplying both sides by the least common denominator of all fractions appearing in the equation, namely $4x$: $x + 4 = 2x$. Now, subtract x from both sides: $x = 4$.

So, it takes Anne 4 hours to complete the job on her own.

283. d. Translating the sentence, "The difference between six times the quantity $6x + 1$ minus three times the quantity $x - 1$ is 108," into symbolic form results in the equation: $6(6x + 1) - 3(x - 1) = 108$. Remember to use parentheses for the expression when the words *the quantity* are used. Perform the distributive property twice on the left side of the equation: $36x + 6 - 3x + 3 = 108$. Combine like terms on the left side of the equation: $33x + 9 = 108$. Subtract 9 from both sides of the equation: $33x + 9 - 9 = 108 - 9$. Simplify: $33x = 99$. Divide both sides of the equation by 33: $\frac{33x}{33} = \frac{99}{33}$. The variable is now isolated: $x = 3$.

284. a. This problem translates to the equation $^-4(x + 8) + 6x = 2x + 32$. Remember to use parentheses for the expression when the words *the quantity* are used. Use the distributive property on the left side of the equation: $^-4x - 32 + 6x = 2x + 32$. Combine like terms on the left side of the equation: $2x - 32 = 2x + 32$. Subtract $2x$ from both sides of the equation: $2x - 2x - 32 = 2x - 2x + 32$. The two sides are not equal. There is no solution: $^-32 \neq 32$.

285. d. Since there are 11 inches beyond the base height of 5 feet for a man whose height is 5 feet 11 inches, we see that the lower bound for the range of a healthy man is $130 + 3 \times 11$ pounds, and the upper bound for the range is pounds. So, the range for a healthy man of this height is 163 pounds to 196 pounds.

286. b. Let x = the number of CDs Kathleen has. Four more than one half the number can be written as $\frac{1}{2}x + 4$. Set this amount equal to 16, which is the number of CDs Michael has. To solve this, subtract 4 from both sides of the equation: $\frac{1}{2}x + 4 - 4 = 16 - 4$. Multiply each side of the equation by 2: $\frac{2x}{2} = 2 \times 12$. The variable is now isolated: $x = 24$ CDs.

287. d. Since the perimeter of the square is $x + 4$, and a square has four equal sides, we can use the perimeter formula for a square to find the answer to the question: $P = 4s$ where P = perimeter and s = side length of the square. Substituting the information given in the problem, $P = x + 4$ and $s = 24$, yields the equation: $x + 4 = 4(24)$. Simplifying yields $x + 4 = 96$. Subtract 4 from both sides of the equation: $x + 4 - 4 = 96 - 4$. Simplify: $x = 92$.

288. b. Let x = the width of the rectangle. Let $x + 3$ = the length of the rectangle, since the length is 3 inches greater than the width. Perimeter is the distance around the rectangle. The formula is *Perimeter = length + width + length + width*, $P = l + w + l + w$, or $P = 2l + 2w$. Substitute the *let* statements made previously for l and w and set the perimeter (P) equal to 21 into the formula: $21 = 2(x + 3) + 2(x)$. Use the distributive property on the right side of the equation: $21 = 2x + 6 + 2x$. Combine like terms of the right side of the equation: $21 = 4x + 6$. Subtract 6 from both sides of the equation: $21 - 6 = 4x + 6 - 6$. Simplify: $15 = 4x$. Divide both sides of the equation by 4: $\frac{15}{4} = \frac{4x}{4}$. The variable is now isolated: 3.75 inches $= x$.

289. c. Let x be Bart's score on the fifth exam. Since the fifth exam score counts twice, the average we must compute is for six scores. The expression for the average of the scores is $\frac{80 + 76 + 89 + 69 + x}{6} = \frac{314 + 2x}{6}$.

The smallest value x can be is the value that gives an average of 80. To find this value, we must solve the equation $\frac{314 + 2x}{6} = 80$. To solve this equation, multiply both sides by 6: $314 + 2x = 480$. Next, subtract 314 from both sides: $2x = 166$. Finally, divide both sides by 2: $x = 83$. So, the smallest value x can be is 83.

The largest value x can be is the value that gives an average of 85. To find this value, we must solve the equation $\frac{314 + 2x}{6} = 85$. To solve this equation, multiply both sides by 6: $314 + 2x = 510$. Next, subtract 314 from both sides: $2x = 196$. Finally, divide both sides by 2: $x = 98$. So, the largest value x can be is 98.

Therefore, Bart's score on the fifth exam must be between 83 and 98 to ensure that his exam average falls between 80 and 85.

290. a. Two consecutive even integers are even numbers in order, such as 4 and 6 or ⁻30 and ⁻32, which are each 2 units apart. Let x = the first consecutive even integer. Let $x + 2$ = the second (and larger) consecutive even integer. *Sum* is a key word for addition so the equation becomes $(x) + (x + 2) = 126$. Combine like terms on the left side of the equation: $2x + 2 = 126$. Subtract 2 from both sides of the equation: $2x + 2 - 2 = 126 - 2$; simplify: $2x = 124$. Divide each side of the equation by 2: $\frac{2x}{2} = \frac{124}{2}$. The variable is now isolated: $x = 62$. Therefore the larger integer is: $62 + 2 = 64$.

291. d. The smallest increase (of 10%) on the cost of the laptop amounts to $0.10(\$1,300) = \$130$. Likewise, the largest increase (of 20%) on the cost of the laptop amounts to $0.20(\$1,300) = \$260$. So, during the holidays, the cost of the laptop ranges from $\$1,300 + \$130 = \$1,430$ to $\$1,300 + \$260 = \$1,560$.

292. c. Three consecutive even integers are even numbers in order, like 4, 6, and 8 or ⁻30, ⁻28 and ⁻26 which are each 2 units apart. Let x = the first and smallest consecutive even integer. Let $x + 2$ = the second consecutive even integer. Let $x + 4$ = the third and largest consecutive even integer. *Sum* is a key word for addition, so the equation becomes $(x) + (x + 2) + (x + 4) = 114$. Combine like terms on the left side of the equation: $3x + 6 = 114$. Subtract 6 from both sides of the equation: $3x + 6 - 6 = 114 - 6$; simplify: $3x = 108$. Divide each side of the equation by 3: $\frac{3x}{3} = \frac{108}{3}$. The variable is now isolated: $x = 36$; therefore, the next larger integer is: $36 + 2 = 38$. The largest even integer would be: $36 + 4 = 40$.

293. d. Let t = the amount of time traveled. Using the formula *distance* = *rate* × *time*, substitute the rates (or average speeds) of each car and multiply by t to find the distance traveled by each car. Therefore, $63t$ = distance traveled by one car, and $59t$ = distance traveled by the other car. Since the cars are traveling in opposite directions, the total distance traveled by both cars is the sum of these distances: $63t + 59t$. Set this equal to the total distance of 610 miles: $63t + 59t = 610$. Combine like terms on the left side of the equation: $122t = 610$. Divide each side of the equation by 122: $\frac{122t}{122} = \frac{610}{122}$; the variable is now isolated: $t = 5$. In 5 hours, the two cars will be 610 miles apart.

294. d. Use the formula *distance = rate × time* for each train and add these values together so that the distance equals 822 miles. For the first train, $d = 65t$ and for the second train $d = 72t$, where d is the distance and t is the time in hours. Add the distances and set them equal to 822: $65t + 72t = 822$. Combine like terms on the left side of the equation: $137t = 822$; divide both sides of the equation by 137: $\frac{137t}{137} = \frac{822}{137}$. The variable is now isolated: $t = 6$. In 6 hours, the two trains will be 822 miles apart.

295. d. Use the formula *distance = rate × time* for each train and add these values together so that the distance equals 1,029 miles. For the first train, $d = 45t$ and for the second train, $d = 53t$, where d is the distance and t is the time in hours. Add the distances and set them equal to 1,029: $45t + 53t = 1,029$. Combine like terms on the left side of the equation: $98t = 1,029$; divide both sides of the equation by 98: $\frac{98t}{98} = \frac{1,029}{98}$. The variable is now isolated: $t = 10.5$ hours. The two trains will pass in 10.5 hours.

296. c. Translate the sentence, "Nine minus five times a number, x, is no less than 39," into symbols: $9 - 5x \geq 39$. Subtract 9 from both sides of the inequality: $9 - 9 - 5x \geq 39 - 9$. Simplify: $^-5x \geq 30$; divide both sides of the inequality by $^-5$. Remember that when dividing or multiplying each side of an inequality by a negative number, the inequality symbol changes direction: $\frac{^-5x}{^-5} \leq \frac{^-30}{^-5}$. The variable is now isolated: $x \leq {}^-6$.

297. a. This problem is an example of a compound inequality, where there is more than one inequality in the question. Let $x =$ the total amount of gumdrops Will originally has. Set up the compound inequality, and then solve it as two separate inequalities. The second sentence in the problem can be written as $2 < x - 2 < 6$. The two inequalities are $2 < x - 2$ and $x - 2 < 6$. Add 2 to both sides of both inequalities: $2 + 2 < x - 2 + 2$ and $x - 2 + 2 < 6 + 2$; simplify: $4 < x$ and $x < 8$. If x is greater than four and less than eight, it means that the solution is between 4 and 8. This can be shortened to $4 < x < 8$.

298. a. This inequality shows a solution set where y is greater than or equal to $^-3$ and less than or equal to eight. Both $^-3$ and 8 are in the solution set because of the word *inclusive*, which includes them. This can be represented by the compound inequality $^-3 \leq y \leq 8$.

299. b. Let x = the unknown number. Remember that *quotient* is a key word for division, and *at least* means greater than or equal to. From the question, the sentence would translate to: $\frac{x}{2} + 5 \geq x$. Subtract 5 from both sides of the inequality: $\frac{x}{2} + 5 - 5 \geq x - 5$; simplify: $\frac{x}{2} \geq x - 5$. Multiply both sides of the inequality by 2: $\frac{x}{2} \times 2 \geq (x - 5) \times 2$; simplify: $x \geq (x - 5)2$. Use the distributive property on the right side of the inequality: $x \geq 2x - 10$. Add 10 to both sides of the inequality: $x + 10 \geq 2x - 10 + 10$; simplify: $x + 10 \geq 2x$. Subtract x from both sides of the inequality: $x - x + 10 \geq 2x - x$. The variable is now isolated: $10 \geq x$. The number is at most 10.

300. d. Let x = the number of hours Cindy worked. Let $2x + 3$ = the number of hours Carl worked. Since the total hours added together was at most 48, the inequality would be $(x) + (2x + 3) \leq 48$. Combine like terms on the left side of the inequality: $3x + 3 \leq 48$. Subtract 3 from both sides of the inequality: $3x + 3 - 3 \leq 48 - 3$; simplify: $3x \leq 45$. Divide both sides of the inequality by 3: $\frac{3x}{3} \leq \frac{45}{3}$; the variable is now isolated: $x \leq 15$. The maximum amount of hours Cindy worked was 15.

301. b. Choices **a** and **d** should be omitted because the negative values do not make sense for this problem using time and cost. Choice **b** substituted into the equation, would be $6 = 2(2) + 2$ which simplifies to $6 = 4 + 2$. Thus, $6 = 6$. The coordinates in choice **c** are reversed from choice **b** and will not work if substituted for x and y.

302. a. Let x = the total minutes of the call. Therefore, $x - 1$ = the additional minutes of the call. In order to calculate the cost, the charge is 35 cents plus 15 cents multiplied by the number of additional minutes. If y represents the total cost, then y equals $0.35 plus $0.15 times the quantity $x - 1$. This translates to $y = \$0.35 + \$0.15(x - 1)$, or $y = \$0.15(x - 1) + \$0.35$.

303. d. Let x = the total miles of the ride. Therefore, $x - 1$ = the additional miles of the ride. The correct equation takes $1.25 and adds it to $1.15 multiplied by the number of additional miles. This becomes y (the total cost) $= \$1.25 + \$1.15(x - 1)$, which is the same equation as $y = \$1.15(x - 1) + \$1.25$.

304. c. The total amount will be $4.85 plus $2 multiplied by the number of ounces, x. This translates to $\$4.85 + 2x$, which is the same as $\$2x + \$4.85$. This value needs to be less than or equal to $10, which can be written as $\$2x + \$4.85 \leq 10$.

305. b. Let x = the number of checks written that month. Green Bank's fees would therefore be represented by $0.10x + \$3$ and Savings-R-Us would be represented by $0.05x + \$4.50$. To find the value for which the banks charge the same amount, set the two expressions equal to each other: $0.10x + \$3 = \$0.05x + \$4.50$. Subtract \$3 from both sides: $0.10x + \$3 - \$3 = \$0.05x + \$4.50 - \$3$. This now becomes: $0.10x = \$0.05x + \$1.50$. Subtract $0.05x$ from both sides of the equation: $0.10x - \$0.05x = \$0.05x - \$0.05x + \$1.50$; this simplifies to: $0.05x = \$1.50$. Divide both sides of the equation by 0.05: $\frac{\$0.05x}{\$0.5} = \frac{\$105}{\$0.5}$. The variable is now isolated: $x = 30$. Monthly costs would be the same for both banks if 30 checks were written.

306. d. Let x = the number of miles traveled in the taxi. The expression for the cost of a ride with Easy Rider would be $1.25x + \$2$. The expression for the cost of a ride with Luxury Limo is $1x + \$3.25$. To solve, set the two expressions equal to each other: $1.25x + \$2 = \$1x + 3.25$. Subtract \$2 from both sides: $1.25x + \$2 - \$2 = \$1x + \$3.25 - \$2$. This simplifies to: $1.25x = \$1x + \$1.25$; subtract $1x$ from both sides: $1.25x - \$1x = \$1x - \$1x + 1.25$. Divide both sides of the equation by 0.25: $\frac{\$0.25x}{\$0.25} = \frac{\$1.25}{\$0.25}$. The variable is now isolated: $x = 5$; the cost for both services would be the same if the trip were 5 miles long.

307. b. Let x = the first integer and y = the second integer. The equation for the sum of the two integers is $x + y = 36$, and the equation for the difference between the two integers is $x - y = 6$. To solve these using the elimination method, combine like terms vertically, and the variable of y cancels out.

$$x + y = 36$$
$$\underline{x - y = 6}$$

This results in: $2x = 42$, so $x = 21$.
Substitute the value of x into the first equation to get $21 + y = 36$. Subtract 21 from both sides of this equation to get an answer of $y = 15$.

308. c. Let x = the greater integer and y = the lesser integer. From the first sentence in the question we get the equation $x = y + 2$. From the second sentence in the question we get $y + 2x = 7$. Substitute $x = y + 2$ into the second equation: $y + 2(y + 2) = 7$; use the distributive property to simplify: $y + 2y + 4 = 7$. Combine like terms: $3y + 4 = 7$; subtract 4 from both sides of the equation: $3y + 4 - 4 = 7 - 4$. Simplify: $3y = 3$. Divide both sides of the equation by 3: $\frac{3y}{3} = \frac{3}{3}$; therefore $y = 1$. Since the greater integer is two more than the lesser, the greater is $1 + 2 = 3$.

309. a. Since P represents the estimated value (in dollars) of the house, the range of the selling price is $0.86P$ to $0.95P$. If the house sold for $0.86P$ dollars, then the loss incurred would be $P - 0.86P = 0.14P$. Similarly, if the house sold for $0.95P$ dollars, the loss incurred would be $P - 0.95P = 0.05P$. Thus, the range of the amount of money lost by selling the house in a "buyer's market" is $0.05P$ to $0.14P$.

310. b. Since there is a ±3% margin of error in the thermostat reading, the actual temperature of the house corresponding to a thermostat reading of 72° F is between 72° F $- 0.03(72° F) = 69.84°$ F and and 72° F $= 74.16°$ F.

311. b. Let l = the length of the rectangle and w = the width of the rectangle. Since the width is 6 inches less than 3 times the length, one equation is $w = 3l - 6$. The formula for the perimeter of the rectangle is $2l + 2w = 104$. Substituting the first equation into the perimeter equation for w results in $2l + 2(3l - 6) = 104$. Use the distributive property on the left side of the equation: $2l + 6l - 12 = 104$. Combine like terms on the left side of the equation: $8l - 12 = 104$; add 12 to both sides of the equation: $8l - 12 + 12 = 104 + 12$. Simplify: $8l = 116$. Divide both sides of the equation by 8: $\frac{8l}{8} = \frac{116}{8}$. Therefore, the length is $l = 14.5$ inches, and the width is $w = 3(14.5) - 6 = 37.5$ inches.

312. d. Since there is a 0.1% margin of error in the breathalyzer reading, the lowest actual blood alcohol level that a person can have and still get arrested is $0.07 - 0.001 = 0.069$. Since A stands for the blood alcohol reading for a person, if $A < 0.069$ then the person will not get arrested for driving under the influence.

313. c. Let x = the amount invested at 12% interest and y = the amount invested at 15% interest. Since the amount invested at 15% is $100 more then twice the amount at 12%, then $y = 2x + \$100$. Since the total interest earned was $855, use the equation $0.12x + 0.15y = \$855$. You have two equations with two variables. Use the second equation and substitute $(2x + \$100)$ for y: $0.12x + 0.15(2x + \$100) = \$855$. Use the distributive property: $0.12x + 0.3x + \$15 = \$855$. Combine like terms: $0.42x + \$15 = \$855$. Subtract $\$15$ from both sides: $0.42x + \$15 - \$15 = \$855 - \$15$; simplify: $0.42x = \$840$. Divide both sides by 0.42: $\frac{0.42x}{0.42} = \frac{\$840}{0.42}$. Therefore, $x = \$2,000$, which is the amount invested at 12% interest.

314. **c.** Let $x =$ the amount invested at 8% interest. Since the total interest earned is $\$405.50$, use the equation $0.06(\$4,000) + 0.08x = \$405.50$. Simplify the multiplication: $\$240 + 0.08x = \$405.50$. Subtract $\$240$ from both sides: $\$240 - \$240 + 0.8x = \$405.50 - \$240$; simplify: $0.08x = \$165.50$. Divide both sides by 0.08: $\frac{0.08x}{0.08} = \frac{\$165.50}{0.008}$. Therefore, $x = \$2,068.75$, which is the amount invested at 8% interest.

315. **d.** Let $x =$ the amount of coffee purchased at $\$3$ per pound. Let $y =$ the total amount of coffee purchased. If there are 18 pounds of coffee at $\$2.50$ per pound, then the total amount of coffee purchased can be expressed as $y = x + 18$. Use the equation $\$3x + \$2.50(18) = \$2.85y$ since the average cost of the y pounds of coffee is $\$2.85$ per pound. To solve, substitute $y = x + 18$ into $3x + 2.50(18) = \$2.85y$: $\$3x + \$2.50(18) = \$2.85(x + 18)$. Multiply on the left side and use the distributive property on the right side: $\$3x + \$45 = \$2.85x + \$51.30$. Subtract $\$2.85x$ on both sides: $\$3x - \$2.85x + \$45 = \$2.85x - \$2.85x + \$51.30$. Simplify: $\$0.15x + \$45 = \$51.30$. Subtract $\$45$ from both sides: $\$0.15x + \$45 - \$45 = \$51.30 - \$45$. Simplify: $\$0.15x = \$6.30$. Divide both sides by $\$0.15$: $\frac{\$0.15x}{\$0.15} = \frac{\$6.30}{\$0.15}$; so, $x = 42$ pounds, which is the amount of coffee purchased that costs $\$3$ per pound. Therefore, the total amount of coffee purchased is $42 + 18 = 60$ pounds.

316. **c.** Let $x =$ the amount of candy purchased at $\$1.90$ per pound. Let $y =$ the total number of pounds of candy purchased. If it is known that there are 40 pounds of candy at $\$2.15$ per pound, then the total number of pounds of candy purchased can be expressed as $y = x + 40$. Use the equation $\$1.90x + \$2.15(40) = \$162$ since the total amount of money spent was $\$162$. Multiply on the left side: $\$1.90x + \$86 = \$162$. Subtract $\$86$ from both sides: $\$1.90x + \$86 - \$86 = \$162 - \$86$. Simplify: $\$1.90x = \$76$. Divide both sides by $\$1.90$: $\frac{\$190x}{\$1.90} = \frac{\$76}{\$1.90}$; so, $x = 40$ pounds, which is the amount of candy purchased that costs $\$1.90$ per pound. Therefore, the total amount of candy purchased is $40 + 40 = 80$ pounds.

317. a. Let x = the number of marigold packets purchased at \$1 per packet. Let y = the number of marigold packet purchased at \$1.26 per packet. Since there are 50 more packets of the \$1.26 seeds than the \$1 seeds, $y = x + 50$. Use the equation \$1x + \$1.26y = \$420 to find the total number of packets purchased of each. By substituting into the second equation, you get \$1x + \$1.26(x + 50) = \$402. Multiply on the left side using the distributive property: \$1x + \$1.26x + \$63 = \$402. Combine like terms on the left side: \$2.26x + \$63 = \$402. Subtract \$63 from both sides: \$2.26x + \$63 – \$63 = \$402 – \$63. Simplify: \$2.26x = \$339. Divide both sides by \$2.26: $\frac{\$2.26x}{\$2.26} = \frac{\$3.39}{\$2.26}$; so, x = 150 packets, which is the number of packets purchased that cost \$1 each.

318. a. Let x = the amount of 3% iodine solution used. Let y = the amount of 20% iodine solution used. Since the total amount of solution made was 85 oz., then $x + y = 85$. The amount of each type of solution added together and set equal to the amount of 19% solution can be expressed in the equation $0.03x + 0.20y = 0.19(85)$. Use both equations to solve for x. Multiply the second equation by 100 to eliminate the decimal point: $3x + 20y = 19(85)$. Simplify that equation: $3x + 20y = 1615$. Multiply the first equation by ⁻20: $^-20x + {}^-20y = {}^-1700$. Add the two equations to eliminate y: $^-17x + 0y = {}^-85$. Divide both sides of the equation by ⁻17: $-\frac{17x}{-17} = \frac{-85x}{-17}$; x = 5. The amount of 3% iodine solution is 5 ounces.

319. c. Let x = the amount of 34% acid solution used. Let y = the amount of 18% iodine solution used. Since the total amount of solution was 30 oz., then $x + y = 30$. The amount of each type of solution used added together and set equal to the amount of 28% solution can be expressed in the equation $0.34x + 0.18y = 0.28(30)$. Use both equations to solve for x. Multiply the second equation by 100 to eliminate the decimal point: $34x + 18y = 28(30)$; simplify that equation: $34x + 22y = 840$. Multiply the first equation by ⁻18: $^-18x + {}^-18y = {}^-540$. Add the two equations to eliminate y: $16x + 0y = 300$. Divide both sides of the equation by 16: $\frac{16x}{16} = \frac{300}{16}$, x = 18.75. The amount of 34% acid solution used was 18.75 ounces.

320. b. Let x = Ellen's age and y = Bob's age. Since Bob is 2 years away from being twice as old as Ellen, then $y = 2x – 2$. The sum of twice Bob's age and three times Ellen's age is 66 and gives a second equation of $2y + 3x = 66$. Substituting the first equation for y into the second equation results in $2(2x - 2) + 3x = 66$. Use the distributive property on the left side of the equation: $4x - 4 + 3x = 66$; combine like terms on the left side of the equation: $7x - 4 = 66$. Add 4 to both sides of the equation: $7x - 4 + 4 = 66 + 4$. Simplify: $7x = 70$. Divide both sides of the equation by 7: $\frac{7x}{7} = \frac{70}{7}$. The variable is now isolated: $x = 10$. Therefore, Ellen is 10 years old.

321. **d.** Let x = Shari's age and y = Sam's age. Since Sam's age is 1 year less than twice Shari's age, this gives the equation $y = 2x - 1$. Since the sum of their ages is 92, this gives a second equation of $x + y = 92$. By substituting the first equation into the second for y, this results in the equation $x + (2x - 1) = 92$. Combine like terms on the left side of the equation: $3x - 1 = 92$. Add 1 to both sides of the equation: $3x - 1 + 1 = 92 + 1$. Simplify: $3x = 93$. Divide both sides of the equation by 3: $\frac{3x}{3} = \frac{93}{3}$. The variable is now alone: $x = 31$. Therefore, Shari's age is 31, and Sam's age is $2(31) - 1 = 61$ years old.

322. **c.** Let x = the number of minutes of leisure time Ted has on Saturday. Ted spends $\frac{3}{8}x$ watching TV, $\frac{3}{16}x$ minutes cleaning, $\frac{1}{4}x$,minutes reading a novel, and 30 minutes surfing the Internet. The sum of all this time is x minutes. This results in the following equation: $\frac{3}{8}x + \frac{3}{16}x + \frac{1}{4}x = x$. To solve for x, multiply both sides by the least common denominator of all fractions appearing in the equation, namely 16: $6x + 3x + 4x + 480 = 16x$. Next, simplify the left side: $13x + 480 = 16x$. Next, subtract 13 from both sides: $480 = 3x$. Finally, divide both sides by 3: $x = 160$ minutes of Saturday leisure time.
Therefore, Ted spends $\frac{3}{8}(160) = 60$ minutes watching TV.

323. **a.** Let d be Danielle's age. Then, Mary's age is $2d - 5$, Karen's age is $1 + (2d - 5)$, and Sue's age is $3[1 + (2d - 5)]$. Simplifying, we conclude that Sue's age in terms of d (which is Danielle's age) is $3[2d - 4] = 6d - 12$.

324. **b.** Let x = the price of a student ticket and y = the price of an adult ticket. The first sentence, "The price of a student ticket is \$1 more than half of an adult ticket" gives the equation $x = \frac{1}{2}y + \$1$; the second sentence, "six adult tickets and four student tickets cost \$76," gives the equation $6y + 4x = \$76$. Substitute the first equation into the second for x: $6y + 4(\frac{1}{2}y + \$1) = \$76$. Use the distributive property on the left side of the equation: $6y + 2y + \$4 = \$76$. Combine like terms: $8y + \$4 = \$76$. Subtract \$4 on both sides of the equation: $8y + \$4 - \$4 = \$76 - \$4$; simplify: $8y = \$72$. Divide both sides by 8: $\frac{8y}{8} = \frac{\$72}{8}$. The variable is now isolated: $y = \$9$. The cost of one adult ticket is \$9.

325. c. Since there are 32 MP3 downloads, the first 10 are each purchased at x dollars each, which results in the cost $10x$ dollars. Also, the next 16 are each purchased at $\frac{x}{2}$ dollars each, which results in the cost $16(\frac{x}{2})$ dollars. The number of MP3 downloads beyond the initial 26 is $32 - 26 = 6$; each of these is purchased at $\frac{x}{4}$ dollars each, which results in a cost of $6(\frac{x}{4})$ dollars. The sum of these three costs is the total cost: $10x + 16(\frac{x}{2}) + 6(\frac{x}{4}) = 10x + 8x + \frac{3x}{2} = 18x + 1.5x = 19.5x$ dollars.

326. c. The terms $3x$ and $5x$ are like terms because they have exactly the same variable with the same exponent. Therefore, you just add the coefficients and keep the variable: $3x + 5x = 8x$ miles.

327. b. Let x be the number of nickels in the piggy bank. Then, there are $65 - x$ dimes in the bank. Multiply the number of nickels by the value of a single nickel ($0.05) and the number of dimes by the value of a single dime ($0.10) to get the portion of the total $5.00 that the nickels and dimes each contributes. The value contributed by the nickels is $0.05x$, and the value contributed by the dimes is $0.10(65 - x)$. The sum of these values is $5.00: $0.05x + $0.10(65 - x) = 5.00. To solve for x, multiply both sides by 100 to clear the decimals: $5x + $10(65 - x) = 500. Next, simplify the left side: $5x + $650 - $10x = 500, which further simplifies to $650 - $5x = 500. Next, add $5x$ to both sides and subtract $500 from both sides: $5x = 150. Finally, divide both sides by $5: $x = 30$ nickels. Therefore, there are $65 - 30 - 35$ dimes.

328. b. Since the area of a rectangle is *Area = length × width*, multiply $(x^3)(x^4)$. When multiplying like bases, add the exponents: $x^{3+4} = x^7$.

329. c. The only pair of dimensions that satisfies the equation is the one given in choice **c**. Substituting $b = 3$ into the equation yields h $= (\frac{1}{2}(3) + 2)^2 = (\frac{3}{2} + 2)^2 = (\frac{7}{2})^2 = \frac{49}{4} = 12\frac{1}{4} = 12.25$.

330. a. Since the area of a parallelogram is *Area = base × height*, then the area divided by the base would give you the height: $\frac{x^8}{x^4}$. When dividing like bases, subtract the exponents: $x^{8-4} = x^4$.

331. c. We must solve the equation $900 - $20m = 300 for m. To do so, add $20m$ and subtract $300 from both sides: $20m = 600. Finally, divide both sides by $20: $m = 30$ months.

332. **a.** The numerical translation of the question is $\frac{6x^2 \cdot 4xy^2}{3x^3y}$. The key word *product* tells you to multiply $6x^2$ and $4xy^2$. The result is then divided by $3x^3y$. Use the commutative property in the numerator to arrange like variables and the coefficients together: $\frac{6 \times 4x^2y^2}{3x^3y}$. Multiply in the numerator. Remember that $x^2 \cdot x = x^2 \cdot x^1 = x^{2+1} = x^3$: $\frac{24x^2y^2}{3x^3y}$. Divide the coefficients: $24 \div 3 = 8$: $\frac{8x^3y^2}{3x^3y}$. Divide the variables by subtracting the exponents: $8x^{3-3}y^{2-1}$; simplify. Recall that any quantity raised to the zero power is equal to 1: $8x^0y^1 = 8y$.

333. **b.** Since the formula for the area of a square is $A = s^2$, substitute a^2b^3 in for s to get $A = (a^2b^3)^2$. Multiply the outer exponent by each exponent inside the parentheses: $a^{2 \times 2}b^{3 \times 2} = a^4b^6$.

334. **a.** The statement in the question would translate to $3x^2(2x^3y)^4$. The word *quantity* reminds you to put that part of the expression in parentheses. Evaluate the exponent by multiplying each number or variable inside the parentheses by the exponent outside the parentheses: $3x^2(2^4x^{3 \times 4}y^4)$; simplify: $3x^2(16x^{12}y^4)$. Multiply the coefficients and add the exponents of like variables: $3(16x^{2+12}y^4)$; simplify: $48x^{14}y^4$.

335. **d.** Since the area of a rectangle is *Area = length × width*, multiply the dimensions to find the area: $2x(4x + 5)$. Use the distributive property to multiply each term inside the parentheses by $2x$: $(2x \times 4x) + (2x \times 5)$. Simplify by multiplying the coefficients of each term and adding the exponents of the like variables: $8x^2 + 10x$.

336. **d.** Substitute $m = 15$ into the given equation to determine the value of the washing machine at 15 months: $V = \$1,200 - \$20(15) = \$1,200 - \$300 = \$900$.

337. **c.** The two numbers in terms of x are $x + 3$ and $x + 4$ since *increased by* tells you to add. *Product* tells you to multiply these two quantities: $(x + 3)$ $(x + 4)$. Use **FOIL** (**F**irst terms of each binomial multiplied, **O**uter terms of each multiplied, **I**nner terms of each multiplied, and **L**ast term of

each binomial multiplied) to multiply the binomials:
$(x \cdot x) + (4 \cdot x) + (3 \cdot x) + (3 \cdot 4)$;
simplify each term: $x^2 + 4x + 3x + 12$.
Combine like terms: $x^2 + 7x + 12$.

338. **d.** Since the area of a rectangle is *Area = length × width*, multiply the two
expressions together: $(2x - 1)(x + 6)$. Use **FOIL** (**F**irst terms of each
binomial multiplied, **O**uter terms of each multiplied, **I**nner terms of
each multiplied, and **L**ast term of each binomial multiplied) to multiply
the binomials: $(2x \cdot x) + (2x \cdot 6) - (1 \cdot x) - (1 \cdot 6)$.
Simplify: $2x^2 + 12x - x - 6$; combine like terms: $2x^2 + 11x - 6$.

339. **a.** Use the formula *distance = rate × time*. By substitution,
distance $= (4x^2 - 2) \times (3x - 8)$. Use **FOIL** (**F**irst terms of each
binomial multiplied, **O**uter terms of each multiplied, **I**nner terms
of each multiplied, and **L**ast term of each binomial multiplied) to
multiply the binomials:
$(4x^2 \cdot 3x) - (8 \cdot 4x^2) - (2 \cdot 3x) - (2 \cdot {}^-8)$.
Simplify each term: $12x^3 - 32x^2 - 6x + 16$.

340. **d.** Since the formula for the volume of a prism is $V = Bh$, where B is the
area of the base and h is the height of the prism, $V = (x - 3)(x^2 + 4x + 1)$.
Use the distributive property to multiply the first term of the binomial,
x, by each term of the trinomial, and then the second term of the binomial, $^-3$, by each term of the trinomial: $x(x^2 + 4x + 1) - 3(x^2 + 4x + 1)$. Then
distribute: $(x \cdot x^2) + (x \cdot 4x) + (x \cdot 1) - (3 \cdot x^2) - (3 \cdot 4x) - (3 \cdot 1)$. Simplify
by multiplying within each term: $x^3 + 4x^2 + x - 3x^2 - 12x - 3$. Use the
commutative property to arrange like terms next to each other. Remember that $1x = x$: $x^3 + 4x^2 - 3x^2 + x\ ^-12x - 3$; combine like terms:
$x^3 + x^2 - 11x - 3$.

341. b. Since the formula for the volume of a rectangular prism is $V = l \times w \times h$, multiply the dimensions together: $(x + 1)(x - 2)(x + 4)$. Use **FOIL** (**F**irst terms of each binomial multiplied, **O**uter terms of each multiplied, **I**nner terms of each multiplied, and **L**ast term of each binomial multiplied) to multiply the first two binomials, $(x + 1)(x - 2)$: $(x \cdot x) + x(-2) + (1 \cdot x) + 1(-2)$. Simplify by multiplying within each term: $x^2 - 2x + x - 2$; combine like terms: $x^2 - x - 2$. Multiply the third factor by this result: $(x + 4)(x^2 - x - 2)$. To do this, use the distributive property to multiply the first term of the binomial, x, by each term of the trinomial, and then the second term of the binomial, 4, by each term of the trinomial: $x(x^2 - x - 2) + 4(x^2 - x - 2)$. Distribute: $(x \cdot x^2) + (x \cdot -x) + (x \cdot {}^-2) + (4 \cdot x^2) + (4 \cdot -x) + (4 \cdot {}^-2)$. Simplify by multiplying in each term: $x^3 - x^2 - 2x + 4x^2 - 4x - 8$. Use the commutative property to arrange like terms next to each other: $x^3 - x^2 + 4x^2 - 2x - 4x - 8$; combine like terms: $x^3 + 3x^2 - 6x - 8$.

342. c. Since the area of a rectangle is found by multiplying length by width, we need to find the factors whose product is $x^2 - 36$. Because x^2 and 36 are both perfect squares ($x^2 = x \cdot x$ and $36 = 6 \cdot 6$), the product, $x^2 - 36$, is called a difference of two perfect squares, and its factors are the sum and difference of the square roots of its terms. Therefore, because the square root of $x^2 = x$ and the square root of $36 = 6$, $x^2 - 36 = (x + 6)(x - 6)$.

343. b. To find the base and the height of the parallelogram, find the factors of this binomial. First look for factors that both terms have in common; $2x^2$ and $10x$ both have a factor of 2 and x. Factor out the greatest common factor, $2x$, from each term: $2x^2 - 10x = 2x(x - 5)$. To check an answer like this, multiply through using the distributive property: $2x(x {}^-5)$; $(2x \cdot x) - (2x \cdot 5)$; simplify: $2x^2 - 10x$.

344. d. Since the formula for the area of a rectangle is *Area = length × width*, find the two factors of $x^2 + 2x + 1$ to get the dimensions. First check to see if there is a common factor in each of the terms or if it is the difference between two perfect squares, but it is neither of these. The next step would be to factor the trinomial into two binomials. To do this, you will be using a method that resembles **FOIL** (**F**irst terms of each binomial multiplied, **O**uter terms of each multiplied, **I**nner terms of each multiplied, and **L**ast term of each binomial multiplied) backwards.

First results in x^2, so the first terms must be $(x)(x)$; **O**uter added to the **I**nner combines to $2x$, and the **L**ast is 1, so you need to find two numbers that add to $+2$ and multiply to $+1$. These two numbers would have to be $+1$ and $+1$: $(x + 1)(x + 1)$. Since the factors of the trinomial are the same, this is an example of a perfect square trinomial, meaning that the farmer's rectangular field is, more specifically, a square field. To check to make sure these are the factors, multiply them by using **FOIL**: $(x \cdot x) + (1 \cdot x) + (1 \cdot x) + (1 \cdot 1)$; multiply in each term: $x^2 + 1x + 1x + 1$; combine like terms: $x^2 + 2x + 1$.

345. a. Let x be the number of girls attending the camp. Then, $540 - x$ boys are attending the camp. Since the ratio of girls to boys is 5 to 7, set up the following proportion to determine x: $\frac{5}{7} = \frac{x}{540 - x}$. To solve for x, cross-multiply: $5(540 - x) = 7x$. Next, simplify the left side: $2,700 - 5x = 7x$. Next, add $5x$ to both sides: $2,700 = 12x$. Finally, divide both sides by 12: $x = 225$. So, there are 225 girls and $540 - 225 = 315$ boys. Finally, to find how many more boys there are than girls attending the camp, subtract: $315 - 225 = 90$.

346. b. Let x be the number of nickels in Jake's collection. Then, there are $x + 3$ dimes and $2(x + 3) = 2x + 6$ quarters in the pile of coins. To obtain the total value of the pile of coins, multiply each of these numbers by the value of each coin. The portion of the total value contributed by the nickels is $\$0.05x$, the portion contributed by the dimes is $\$0.10(x + 3)$, and the portion contributed by the quarters is $\$0.25(2x + 6)$. So, the total value of the pile of coins is $\$0.05x + \$0.10(x + 3) + \$0.25(2x + 6)$.

347. d. Since each roll uses 15 square inches of cardboard, 24 rolls use $24 \times 15 = 360$ square inches of cardboard. So, y mega-packs, each of which contains 24 rolls, uses $360y$ square inches of cardboard.

348. a. To approximate the number of featured stories in next month's issue, we multiply the number of stories in the present month's issue by 1.1. We apply this process y times in succession to determine the number of stories in the issue y months from now. This yields $\underbrace{1.1 \times \ldots \times 1.1}_{y \text{ times}} \times 15 = (1.1)^y \times 15$ featured stories.

349. **c.** Starting with 5,000 tickets, subtract m tickets for each of the 14 days in the first two weeks of sales, and then subtract $2m$ tickets each day for the remaining 7 days to determine the number of tickets available on Day 21. This yields the expression $5,000 - 14m - 7(2m) = 5,000 - 14m - 14m = 5,000 - 28m$ tickets.

350. **d.** In order to convert this number into standard notation, multiply 5.3 by the factor of 10^{-6}. Since 10^{-6} is equal to 0.000001, 5.3×0.000001 is equal to 0.0000053. Equivalently, move the decimal point in 5.3 six places to the left since the exponent on the 10 is -6.

351. **b.** We must determine the value of g that makes $P = 0$. This amounts to solving the equation $5g^2 - 1,125 = 0$. To do so, first add 1,125 to both sides of the equation: $5g^2 = 1,125$. Next, divide both sides by 5: $g^2 = 225$. Finally, note that $15^2 = 225$, so the equation is satisfied when $g = 15$. Therefore, the local hardware store breaks even once it has sold 15 grills.

352. **d.** Let x = the number. The statement, "The square of a number added to 25 equals 10 times the number," translates into the equation $x^2 + 25 = 10x$. Put the equation in the standard form $ax^2 + bx + c = 0$: $x^2 - 10x + 25 = 0$. Factor the left side of the equation: $(x - 5)(x - 5) = 0$. Set each factor equal to zero and solve: $x - 5 = 0$ or $x - 5 = 0$; $x = 5$ or $x = 5$. The number is 5.

353. **b.** Let x = the number. The statement, "The sum of the square of a number and 12 times the number is -27," translates into the equation $x^2 + 12x = -27$. Put the equation in the standard form $ax^2 + bx + c = 0$: $x^2 + 12x + 27 = 0$. Factor the left side of the equation: $(x + 3)(x + 9) = 0$. Set each factor equal to zero and solve: $x + 3 = 0$ or $x + 9 = 0$; $x = -3$ or $x = -9$. The possible values of this number are -3 or -9, the smaller of which is -9.

354. **c.** Since 1 hour = 60 minutes, it follows that x hours = $60x$ minutes. Further, 3 minutes 30 seconds = 3.5 minutes. So, the average number of MP3s Scott listens to in x hours is given by the quotient $\frac{60x}{3.5}$.

355. **c.** Let x be the number of $5 bills the Girl Scout has. Then, she has $(x - 2)$ $10 bills and $(4x)$ $1 bills. The total value of the collection of bills is obtained by multiplying each of these expression by the value of each bill, and then adding these values. The portion of the total value contributed by the $1 bills is $1(4x)$, the portion contributed by the $5 bills is $5(x)$, and the portion contributed by the $10 bills is $10(x - 2)$. So, the total value of the three types of bills is $4x + $5x + $10(x - 2)$. We are told that this value is $170. Equating the two expressions yields the following equation: $4x + $5x + $10(x - 2) = 170. To solve for x, simplify the left side: $9x + $10x - $20 = 170, which is further equivalent to $19x - $20 = 170. Next, add $20 to both sides: $19x = 190. Finally, divide both sides by $19: $x = 10$. The Girl Scout has 10 $5 bills; therefore, she has $10(4) = 40$ $1 bills.

356. **d.** Substitute $s = 2$, $l = 65$, and $w = 35$ into the given formula to determine the number of 5-gallon barrels needed to apply *one* coat of paint: $[8(2)(35 + 65) - 6(2)] \div 397 = 1588 \div 397 = 4$. Next, multiply this number by 2 to obtain the number of 5-gallon barrels of paint needed for two coats: $4 \times 2 = 8$. Finally, since each barrel contains 5 gallons of paint, we conclude that $8 \times 5 = 40$ gallons of paint are needed.

357. **a.** Working together, the two workers can stuff $E + 3E = 4E$ envelopes per minute. Let x be the number of minutes it takes the two workers, working together, to stuff 5,000 envelopes. Set up the following proportion: $\frac{4E \text{ envelopes}}{1 \text{ minute}} = \frac{5,000 \text{ envelopes}}{x \text{ minute}}$. To solve for x, cross multiply: $4E(x) = 5,000$. Then, divide both sides by $4E$: $x = \frac{5,000}{4E} = \frac{1,250}{E}$ minutes.

358. **c.** Let $x =$ the length of the first piece of rope. Then, $2x - 1 =$ the length of the second piece, and $3(2x - 1) + 10 =$ the length of the third piece (60 feet). If we add the lengths of these three smaller pieces, we should get the length of the original piece of rope. This yields the following equation: $x + (2x - 1) + 3(2x - 1) + 10 = 60$. To solve for x, simplify the left side: $9x + 6 = 60$. Next, subtract 6 from both sides: $9x = 54$. Finally, divide both sides by 6: $x = 6$. Therefore, the length of the first piece is 6 feet, the second piece has a length of 11 feet, and the third piece is 43 feet long. So, we conclude that the longest piece of rope is 43 feet long.

359. b. Let x = the number of canisters of Wilson balls purchased. Then, $x + 1$ = the number of canisters of Penn balls purchased. If you multiply the price of one canister of Wilson balls by the number of canisters of Wilson balls, you get the portion of the total amount spent on Wilson balls. The same is true for the Penn balls. This yields the equation \$3.50x + \$2.75(x + 1) = \$40.25. To solve for x, simplify the left side: \$6.25x + \$2.75 = \$40.25. Next, subtract \$2.75 from both sides: \$6.25x = \$37.50. Finally, divide both sides by \$6.25: x = 6. Therefore, the tennis coach bought 6 canisters of Wilson balls and 7 canisters of Penn balls.

360. a. Let x = the number of gallons needed of the 30% nitrogen. Then, since the Payload Specialist is supposed to end up with 10 gallons of the 70% solution, $10 - x$ gallons of the 90% nitrogen are needed. If you multiply the number of gallons of 30% nitrogen by its concentration, you get the amount of nitrogen contained within the 30% solution. This is also true for the 90% nitrogen, as well as for the final 70% solution. This yields the equation $0.30x + 0.90(10 - x) = 0.70(10)$. To solve for x, we first multiply both sides by 100 to clear the decimals: $30x + 90(10 - x) = 70(10)$. Next, simplify the left side: $30x + 900 - 90x = 700$, which is equivalent to $900 - 60x = 700$. Next, add $60x$ and subtract 700 from both sides: $60x = 200$. Finally, divide both sides by 60: $x = \frac{200}{60} = 3\frac{1}{3}$. So, the Payload Specialist should mix $3\frac{1}{3}$ gallons of the 30% nitrogen solution with $10 - 3\frac{1}{3} = 6\frac{2}{3}$ gallons of the 90% nitrogen solution to obtain 10 gallons of the desired 70% solution.

361. b. Let x = the lesser integer and $x + 1$ = the greater integer. Since *product* is a key word for multiplication, the equation is $x(x + 1) = 90$. Multiply using the distributive property on the left side of the equation: $x^2 + x = 90$. Put the equation in standard form $ax^2 + bx + c = 0$: $x^2 + x - 90 = 0$. Factor the trinomial: $(x - 9)(x + 10) = 0$. Set each factor equal to zero and solve: $x - 9 = 0$ or $x + 10 = 0$; $x = 9$ or $x = {}^-10$. Since you are looking for a positive integer, reject the x-value of $^-10$. Therefore, the lesser positive integer would be 9.

362. b. Let x = the page number of the faster reader. Then the slower reader is on page $x - 4$. The sum of the two page numbers is given to be 408. This yields the equation $x + (x - 4) = 408$. To solve for x, simplify the left side: $2x - 4 = 408$. Next, add 4 to both sides: $2x = 412$. Finally, divide both sides by 2: $x = 206$. So, the slower brother is on page $206 - 4 = 202$.

363. a. We use the formula *distance = rate × time* to determine the distance traveled by each of the two ships.
Eastbound Ship: Let x be the speed of this ship (in miles per hour). It travels for 3.5 hours, the distance it travels is $3.5x$ miles.
Westbound Ship: The speed of this ship is $x - 10$ (in miles per hour). It also travels for 3.5 hours, the distance it travels is $3.5(x - 10)$ miles. The sum of the distances traveled by these two ships is given to be 175 miles. This yields the following equation:
$3.5x + 3.5(x - 10) = 175$. To solve for x, simplify the left side: $3.5x + 3.5x - 35 = 175$, which is equivalent to $7x - 35 = 175$. Next, add 35 to both sides: $7x = 210$. Finally, divide both sides by 7: $x = 30$ miles per hour. Therefore, the eastbound ship traveled $3.5 \times 30 = 105$ miles.

364. c. Let x = Mike's score on the first exam. Then, his score on the second exam is $x + 10$, and his score on the third exam is $x + 10 + 4 = x + 14$. We are told that the average of the three exam scores is 88. This leads to the following equation: $\frac{x + (x + 10) + (x + 14)}{3} = 88$. To solve for x, first cross multiply: $x + (x + 10) + (x + 14) = 3 \times 88$. Next, simplify both sides of the equation: $3x + 24 = 264$. Next, subtract 24 from both sides: $3x = 240$. Finally, divide both sides by 3: $x = 80$. Substituting this into the expressions for the three exam scores, we conclude that the three exam scores are 80, 90, and 94.

365. c. Let x = the lesser odd integer and $x + 2$ = the greater odd integer. The numerical translation of the sentence, "The sum of the squares of two consecutive odd integers is 74," is the equation $x^2 + (x + 2)^2 = 74$. Multiply $(x + 2)^2$ out as $(x + 2)(x + 2)$ using the distributive property: $x^2 + (x^2 + 2x + 2x + 4) = 74$. Combine like terms on the left side of the equation: $2x^2 + 4x + 4 = 74$. Put the equation in standard form by subtracting 74 from both sides: $2x^2 + 4x - 70 = 0$; factor the trinomial completely: $2(x^2 + 2x - 35) = 0$; $2(x - 5)(x + 7) = 0$. Set each factor equal to zero and solve: $2 \neq 0$ or $x - 5 = 0$ or $x + 7 = 0$; $x = 5$ or $x = {}^-7$. Since you are looking for a positive integer, reject the solution of $x = {}^-7$. Therefore, the smaller positive integer is 5.

366. a. Let x = the lesser integer and $x + 1$ = the greater integer. The sentence, "the difference between the squares of two consecutive integers is 15," can be translated into the equation $(x + 1)^2 - x^2 = 15$. Multiply the binomial $(x + 1)^2$ as $(x + 1)(x + 1)$ using the distributive property: $x^2 + x + x + 1 - x^2 = 15$. Combine like terms: $2x + 1 = 15$; subtract 1 from both sides of the equation: $2x + 1 - 1 = 15 - 1$. Divide both sides by 2: $\frac{2x}{2} = \frac{14}{2}$. The variable is now isolated: $x = 7$. Therefore, the larger consecutive integer is $x + 1 = 8$.

367. c. Let x = the lesser integer and $x + 1$ = the greater integer. The sentence, "The square of one integer is 55 less than the square of the next consecutive integer," can be translated into the equation $x^2 = (x + 1)^2 - 55$. Multiply the binomial $(x + 1)^2$ as $(x + 1)(x + 1)$ using the distributive property: $x^2 = x^2 + x + x + 1 - 55$. Combine like terms: $x^2 = x^2 + 2x - 54$. Subtract x^2 from both sides of the equation: $x^2 - x^2 = x^2 - x^2 + 2x - 54$. Add 54 to both sides of the equation: $0 + 54 = 2x - 54 + 54$. Divide both sides by 2: $\frac{54}{2} = \frac{2x}{2}$. The variable is now isolated: $27 = x$. The value of the lesser integer is 27.

368. c. Let x = the amount each side of the photo is increased. Then, $x + 4$ = the new width and $x + 6$ = the new length. Since $Area = length \times width$, the formula using the new area is $(x + 4)(x + 6) = 168$. Multiply using the distributive property on the left side of the equation: $x^2 + 6x + 4x + 24 = 168$; combine like terms: $x^2 + 10x + 24 = 168$. Subtract 168 from both sides: $x^2 + 10x + 24 - 168 = 168 - 168$. Simplify: $x^2 + 10x - 144 = 0$. Factor the trinomial: $(x - 8)(x + 18) = 0$. Set each factor equal to zero and solve: $x - 8 = 0$ or $x + 18 = 0$; $x = 8$ or $x = {}^-18$. Reject the negative solution because you can't have a negative dimension. Therefore, each dimension of the photo is increased by 8 inches.

369. a. Let x = the amount of reduction. Then, $4 - x$ = the width of the reduced picture and $6 - x$ = the length of the reduced picture. Since $Area = length \times width$, and one-third of the old area of 24 is 8, the equation for the area of the reduced picture is $(4 - x)(6 - x) = 8$. Multiply the binomials using the distributive property: $24 - 4x - 6x + x^2 = 8$; combine like terms: $24 - 10x + x^2 = 8$. Subtract 8 from both sides: $24 - 8 - 10x + x^2 = 8 - 8$. Simplify and place in standard form: $x^2 - 10x + 16 = 0$. Factor the trinomial into 2 binomials: $(x - 2)(x - 8) = 0$. Set each factor equal to zero and solve: $x - 2 = 0$ or $x - 8 = 0$; $x = 2$ or $x = 8$. The solution of 8 is not reasonable because it is greater than the original dimensions of the picture. Accept the solution of $x = 2$, and the smaller dimension of the reduced picture would be $4 - 2 = 2$ inches.

370. d. Let x = the amount that each side of the garden is increased. Then, $x + 20$ = the new width and $x + 24$ = the new length. Since *Area = length × width*, then the area of the old garden is $20 \times 24 = 480$ square feet, and the new area is $480 + 141 = 621$ square feet. The equation using the new area becomes $(x + 20)(x + 24) = 621$. Multiply using the distributive property on the left side of the equation: $x^2 + 24x + 20x + 480 = 621$; combine like terms: $x^2 + 44x + 480 = 621$. Subtract 621 from both sides: $x^2 + 44x + 480 - 621 = 621 - 621$; simplify: $x^2 + 44x - 141 = 0$. Factor the trinomial: $(x - 3)(x + 47) = 0$. Set each factor equal to zero and solve: $x - 3 = 0$ or $x + 47 = 0$; $x = 3$ or $x = -47$. Reject the negative solution because you can't have a negative increase in length. Thus, the length of each side of the garden will be increased by 3 feet, and the new length would be $24 + 3 = 27$ feet.

371. b. Let x = the number of hours it takes Ian and Jack to remodel the kitchen if they are working together. Since it takes Ian 20 hours if working alone, he will complete $\frac{1}{20}$ of the job in one hour, even when he's working with Jack. Similarly, since it takes Jack 15 hours to remodel a kitchen, he will complete $\frac{1}{15}$ of the job in one hour, even when he's working with Ian. Since it takes x hours for Ian and Jack to complete the job together, it stands to reason that at the end of one hour, their combined effort will have completed $\frac{1}{x}$ of the job. Therefore, *Ian's work + Jack's work = combined work*, and we have the equation: $\frac{1}{20} + \frac{1}{15} = \frac{1}{x}$. Multiply through by the least common denominator of 20, 15, and x, which is $60x$: $(60x)(\frac{1}{20}) + (60x)(\frac{1}{15}) = (60x)(\frac{1}{x})$. Simplify: $3x + 4x = 60$. Simplify: $7x = 60$. Divide by 7: $\frac{7x}{7} = \frac{60}{7}$; $x = \frac{60}{7}$, which is about 8.6 hours.

372. a. Let x = the number of hours it takes Peter and Joe to paint a room if they are working together. Since it takes Peter 1.5 hours if working alone, he will complete $\frac{1}{1.5}$ of the job in one hour, even when he's working with Joe. Similarly, since it takes Joe 2 hours to paint a room working alone, he will complete $\frac{1}{2}$ of the job in one hour, even when working with Peter. Since it takes x hours for Peter and Joe to complete the job together, it stands to reason that at the end of one hour, their combined effort will have completed $\frac{1}{x}$ of the job. Therefore, *Peter's work + Joe's work = combined work*, and we have the equation: $\frac{1}{1.5} + \frac{1}{2} = \frac{1}{x}$. Multiply through by the least common denominator of 1.5, 2, and x, which is $6x$: $(6x)(\frac{1}{1.5}) + (6x)(\frac{1}{2}) = (6x)(\frac{1}{x})$. Simplify: $4x + 3x = 6$. Simplify: $7x = 6$. Divide by 7: $\frac{7x}{7} = \frac{6}{7}$; $x = \frac{6}{7}$ hours. Change hours into minutes by multiplying by 60 since there are 60 minutes in one hour. $(60)(\frac{6}{7}) = \frac{360}{7} = 51.42$ minutes, which rounds to 51 minutes.

373. c. Let x = the number of hours it takes Carla and Charles to plant a garden if they are working together. Since it takes Carla 3 hours if working alone, she will complete $\frac{1}{3}$ of the job in one hour, even when she's working with Charles. Similarly, since it takes Charles 4.5 hours to plant a garden working alone, he will complete $\frac{1}{4.5}$ of the job in one hour, even when working with Carla. Since it takes x hours for Carla and Charles to complete the job together, it stands to reason that at the end of one hour, their combined effort will have completed $\frac{1}{x}$ of the job. Therefore, *Carla's work + Charles's work = combined work*, and we have the equation: $\frac{1}{3} + \frac{1}{4.5} = \frac{1}{x}$. Multiply through by the least common denominator of 3, 4.5, and x, which is $9x$: $(9x)(\frac{1}{3}) + (9x)(\frac{1}{4.5}) = (9x)(\frac{1}{x})$. Simplify: $3x + 2x = 9$. Simplify: $5x = 9$. Divide by 5: $\frac{5x}{5} = \frac{9}{5}$; $x = \frac{9}{5}$ hours, which is equal to 1.8 hours.

374. **c.** Let x = the number of hours it will take Jerry to do the job alone. In 1 hour Jim can do $\frac{1}{10}$ of the work, and Jerry can do $\frac{1}{x}$ of the work. As an equation this looks like $\frac{1}{10} + \frac{1}{x} = \frac{1}{4}$, where $\frac{1}{4}$ represents what part of the job they can complete in one hour together. Multiplying both sides of the equation by the least common denominator, $40x$, results in the equation: $4x + 40 = 10x$. Subtract $4x$ from both sides of the equation: $4x - 4x + 40 = 10x - 4x$. This simplifies to $40 = 6x$. Divide each side of the equation by 6: $\frac{40}{6} = \frac{6x}{6}$. Therefore, $6.666 = x$, and it would take Jerry about 6.7 hours to complete the job alone.

375. **d.** Let x = the number of hours Ben takes to clean the garage by himself. In 1 hour Ben can do $\frac{1}{x}$ of the work and Bill can do $\frac{1}{15}$ of the work. As an equation this looks like $\frac{1}{x} + \frac{1}{15} = \frac{1}{6}$, where $\frac{1}{6}$ represents what part they can clean in one hour together. Multiply both sides of the equation by the least common denominator, $30x$, to get the equation $30 + 2x = 5x$. Subtract $2x$ from both sides of the equation: $30 + 2x - 2x = 5x - 2x$. This simplifies to $30 = 3x$, and dividing both sides by 3 results in $x = 10$ hours.

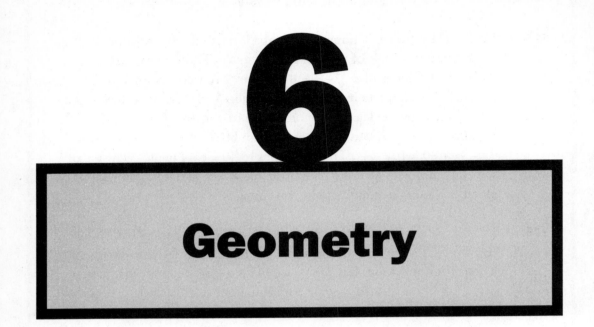

Geometry

The geometry problems in this chapter involve lines, angles, triangles, rectangles, squares, and circles. You will learn how to find length, perimeter, area, circumference, and volume, and how you can apply geometry to everyday situations.

376. Charlie wants to know the area of his rectangular plot of land, which measures 120 ft by 150 ft. Which formula will he use?
    a. $A = s^2$
    b. $A = \pi r^2$
    c. $A = \frac{1}{2}bh$
    d. $A = lw$

377. Dawn wants to compare the volume of a basketball with the volume of a tennis ball. Which formula will she use?
    a. $V = \pi r^2 h$
    b. $V = \frac{4}{3}\pi r^3$
    c. $V = \frac{1}{3}\pi r^2 h$
    d. $V = s^3$

378. Rick is ordering a new triangular sail for his boat. He needs to know the area of the sail. Which formula will he use?

    a. $A = lw$

    b. $A = \frac{1}{2}bh$

    c. $A = bh$

    d. $A = \frac{1}{2}h(b^1 + b^2)$

379. Keith wants to know the surface area of a basketball. Which formula will he use?

    a. $s = 6s^2$

    b. $s = 4\pi r^2$

    c. $s = 2\pi r^2 + 2\pi rh$

    d. $s = \pi r^2 + 2\pi rh$

380. Aaron is installing a ceiling fan in his bedroom. Once the fan is in motion, he needs to know the area the fan will cover. Which formula will he use?

    a. $A = bh$

    b. $A = s^2$

    c. $A = \frac{1}{2}bh$

    d. $A = \pi r^2$

381. Mimi is filling a tennis ball can with water. She wants to know the volume of the cylinder-shaped can. What formula will she use?

    a. $V = \pi r^2 h$

    b. $V = \frac{4}{3}\pi r^3$

    c. $V = \frac{1}{3}\pi r^2 h$

    d. $V = s^3$

382. Audrey is creating a raised flowerbed that is 4.5 ft by 4.5 ft. She needs to calculate how much lumber to buy. If she needs to know the distance around the flowerbed, which formula will she use?

    a. $P = a + b + c$

    b. $A = lw$

    c. $P = 4s$

    d. $C = 2\pi r$

383. Al is painting a right cylinder-shaped storage tank. In order to purchase the correct amount of paint he needs to know the total surface area to be painted. Which formula will he use if he does not paint the bottom of the tank?

a. $S = 2\pi r^2 + 2\pi rh$
b. $S = 4\pi r^2$
c. $S = \pi r^2 + 2\pi rh$
d. $S = 6s^2$

384. Cathy is creating a quilt out of fabric panels that are 6 in by 6 in. She wants to know the total area of her square-shaped quilt. Which formula will she use?

a. $A = s^2$
b. $A = \frac{1}{2}bh$
c. $A = \pi r^2$
d. $A = \frac{1}{2}h(b^1 + b^2)$

385. If Lisa wants to know the distance around her circular table, which has a diameter of 42 in., which formula will she use?

a. $P = 4s$
b. $P = 2l + 2w$
c. $C = \pi d$
d. $P = a + b + c$

386. The area of a rectangle is 24 square inches. The length of the rectangle is 5 inches longer than the width. How many inches is the width?

a. 4
b. 3
c. 6
d. 8

387. To find the volume of a cube that measures 3 cm by 3 cm by 3 cm, which formula would you use?

a. $V = \pi r^2 h$
b. $V = \frac{4}{3}\pi r^3$
c. $V = \frac{1}{3}\pi r^2 h$
d. $V = s3$

388. To find the perimeter of a triangular region, which formula would you use?

    a. $P = a + b + c$

    b. $P = 4s$

    c. $P = 2l + 2w$

    d. $C = 2\pi r$

389. A racquetball court is 40 ft by 20 ft. What is the area of the court?

    a. 60 ft^2

    b. 80 ft^2

    c. 800 ft^2

    d. 120 ft2

390. Allan has been hired to mow the school soccer field, which is 180 ft wide by 330 ft long. If his mower mows strips that are 2 feet wide, how many times must he mow across the width of the lawn?

    a. 90

    b. 165

    c. 255

    d. 60

391. Erin is painting a bathroom with four walls that each measure 9 ft by 5.5 ft. Ignoring the doors, windows and ceiling, what is the area to be painted?

    a. 198 ft^2

    b. 66 ft^2

    c. 49.5 ft^2

    d. 160 ft^2

392. The height of a parallelogram measures 5 meters more than its base. If the area of the parallelogram is 36 square meters, what is the height in meters?

    a. 6

    b. 9

    c. 12

    d. 4

393. A building that is 45 ft tall casts a shadow that is 30 ft long. Nearby, Heather is walking her standard poodle, which casts a shadow that is 2.5 ft long. How tall is Heather's poodle?
   a. 2.75 ft
   b. 3.25 ft
   c. 3.75 ft
   d. 1.67 ft

394. A circular pool is filling with water. Assuming the water level will be 4 ft deep and the diameter is 20 ft when the pool is full, what is the volume of the water needed to fill the pool? (Use $\pi = 3.14$)
   a. 251.2 ft³
   b. 1,256 ft³
   c. 5,024 ft³
   d. 3,140 ft³

395. A cable is attached to a pole 36 ft above ground and fastened to a stake in the ground 15 ft from the base of the pole. In order to keep the pole perpendicular to the ground, how long is the cable?
   a. 22 ft
   b. 39 ft
   c. 30 ft
   d. 51 ft

396. Karen is buying a wallpaper border for her bedroom, which is 12 ft by 13 ft. If the border is sold in rolls of 5 yards each, how many rolls will she need to purchase?
   a. 3
   b. 4
   c. 5
   d. 6

397. Patrick has a rectangular patio with a length that is 5 m less than the diagonal and a width that is 7 m less than the diagonal. If the area of his patio is 195 square meters, what is the length of the diagonal, in meters?
a. 10
b. 8
c. 16
d. 20

398. Samantha owns a rectangular field that has an area of 3,280 square feet. The length of the field is 2 feet more than twice the width. What is the width of the field, in feet?
a. 40
b. 82
c. 41
d. 84

399. The scale on a map shows that 1 inch is equal to 14 miles. Shannon measured the distance of her trip on the map to be 17 inches. How far will she need to travel?
a. 23.8 miles
b. 238 miles
c. 2,380 miles
d. 23,800 miles

400. How far will a bowling ball roll in one rotation if the ball has a diameter of 10 inches? (Use $\pi = 3.14$)
a. 31.4 in
b. 78.5 in
c. 15.7 in
d. 62.8 in

401. A water sprinkler sprays a distance of 10 ft in a circular pattern. What is the circumference of the spray? (Use $\pi = 3.14$)
a. 31.4 ft
b. 314 ft
c. 62.8 ft
d. 628 ft

402. If a triangular sail has a vertical height of 83 ft and a horizontal length of 40 ft, what is the area of the sail?
a. 1,660 ft²
b. 1,155 ft²
c. 201 ft²
d. 3,320 ft²

403. What is the volume of a ball whose radius is 4 inches? Round to the nearest cubic inch. (Use π = 3.14)
a. 201 in³
b. 268 in³
c. 804 in³
d. 33 in³

404. If a tabletop has a diameter of 42 in, what is its surface area to the nearest square inch? (Use π = 3.14)
a. 1,384 in²
b. 1,319 in²
c. 1,385 in²
d. 5,539 in²

405. An orange has a radius of 3.25 inches. Find the volume of one orange, to the nearest hundredth of a cubic inch. (Use π = 3.14)
a. 44.22 in³
b. 1,149.76 in³
c. 132.67 in³
d. 143.72 in³

406. A fire and rescue squad places a 15 ft ladder against a burning building. If the bottom of the ladder is 9 ft from the base of the building, how far up the building will the ladder reach?
a. 8 ft
b. 10 ft
c. 12 ft
d. 14 ft

407. Safety deposit boxes can be rented at the bank. The dimensions of a box are 22 in. by 5 in. by 5 in. What is the volume of the box?

    a. 220 in³

    b. 550 in³

    c. 490 in³

    d. 360 in³

408. How many degrees does a minute hand move in 25 minutes?

    a. 25°

    b. 150°

    c. 60°

    d. 175°

409. Two planes leave the airport at the same time. Minutes later, plane A is 70 miles due north of the airport and plane B is 168 miles due east of the airport. How far apart are the two airplanes?

    a. 182 miles

    b. 119 miles

    c. 163.8 miles

    d. 238 miles

410. If the area of a small pizza is 78.5 in², what size pizza box would best fit the small pizza? (Note: Pizza boxes are measured according to the length of one side.)

    a. 12 in

    b. 11 in

    c. 9 in

    d. 10 in

411. Stuckeyburg is a small town in rural America. The map below shows the official city limits. Use the map to approximate the area of the town.

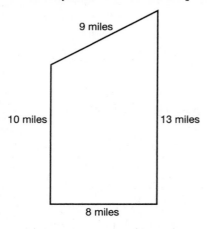

9 miles

10 miles   13 miles

8 miles

    a. 40 miles²
    b. 104 miles²
    c. 93.5 miles²
    d. 92 miles²

412. A rectangular field is to be fenced in completely. The width is 28 yd, and the total area is 1,960 yd². What is the length of the field?
    a. 1,932 yd
    b. 70 yd
    c. 31 yd
    d. 473 yd

413. A circular print is being matted in a square frame. If the frame is 18 in. by 18 in., and the radius of the print is 7 in., what is the area of the matting? (Use π = 3.14)
    a. 477.86 in²
    b. 170.14 in²
    c. 280.04 in²
    d. 288 in²

414. Ribbon is wrapped around a rectangular box that is 10 in by 8 in by 4 in. Using the illustration provided, determine how much ribbon is needed to wrap the box. Assume the amount of ribbon does not include a knot or bow.

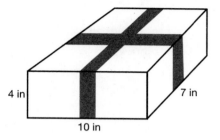

4 in   7 in

10 in

    a. 50 in
    b. 42 in
    c. 22 in
    d. 280 in

415. Pat is making a Christmas tree skirt. She needs to know how much fabric to buy. Using the illustration provided, determine the area of the skirt to the nearest square foot. (Use $\pi = 3.14$)

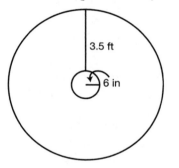

3.5 ft

6 in

    a. 37.7 ft^2
    b. 27 ft^2
    c. 75 ft^2
    d. 38 ft^2

416. Mark intends to tile a kitchen floor, which is 9 ft by 11 ft. How many 6-inch tiles are needed to tile the floor?
    a. 60
    b. 99
    c. 396
    d. 449

417. A framed print measures 36 in by 22 in. If the print is enclosed by a 2-inch matting, what is the length of the diagonal of the print? Round to the nearest tenth of an inch. See the illustration below.

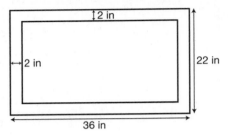

   a. 36.7 in
   b. 39.4 in
   c. 26.5 in
   d. 50 in

418. A garden in the shape of a rectangle is surrounded by a walkway of uniform width. The dimensions of the garden only are 35 feet by 24 feet. The area of the garden and the walkway together is 1,530 square feet. What is the width of the walkway, in feet?
   a. 4
   b. 5
   c. 34.5
   d. 24

419. Barbara is wrapping a wedding gift that is contained within a rectangular box that measures 20 in. by 18 in. by 4 in. How much wrapping paper will she need?
   a. 512 in²
   b. 1,440 in²
   c. 1,024 in²
   d. 92 in²

420. Mark is constructing a walkway around his inground pool. The pool is 20 ft by 40 ft, and the walkway is intended to be 4 ft wide. What is the area of the walkway?
   a. 224 ft²
   b. 416 ft²
   c. 256 ft²
   d. 544 ft²

421. The picture frame shown below has outer dimensions of 8 in by 10 in and inner dimensions of 6 in by 8 in. Find the area of section *A* of the frame.

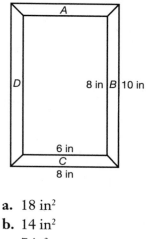

a. 18 in²
b. 14 in²
c. 7 in²
d. 9 in²

For questions 422 and 423, use the following illustration.

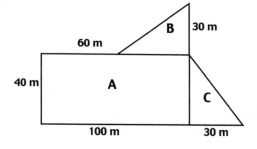

422. John is planning to purchase an irregularly shaped plot of land. Referring to the diagram, find the total area of the land.
a. 6,400 m²
b. 5,200 m²
c. 4,500 m²
d. 4,600 m²

423. Determine the perimeter of the plot of land using the diagram.
   a. 260 m
   b. 340 m
   c. 360 m
   d. 320 m

424. A weather vane is mounted on top of an 18 ft pole. If a 22 ft guy wire is staked to the ground to keep the pole perpendicular, how far is the stake from the base of the pole?
   a. 160 ft
   b. $\sqrt{808}$ ft or $2\sqrt{202}$ ft
   c. 38 ft
   d. $\sqrt{160}$ or $4\sqrt{10}$ ft

425. A surveyor is hired to measure the width of a river. Using the illustration provided, determine the width of the river.

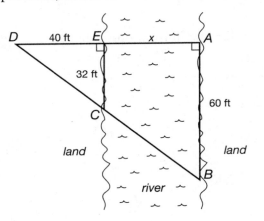

   a. 48 ft
   b. 8 ft
   c. 35 ft
   d. 75 ft

426. A publishing company is designing a book jacket for a newly published textbook. Find the area of the book jacket, given that the front cover is 8 in. wide by 11 in. high, the binding is 1.5 in. by 11 in., and the jacket will extend 2 inches inside the front and rear covers.

   a. 236.5 in²

   b. 192.5 in²

   c. 188 in²

   d. 232 in²

427. A Norman window is to be installed in a new home. Using the dimensions marked on the illustration below, find the area of the window to the nearest tenth of a square inch. (Use $\pi = 3.14$)

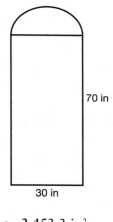

70 in

30 in

   a. 2,453.3 in²

   b. 2,806.5 in²

   c. 147.1 in²

   d. 2,123.6 in²

428. A surveyor is hired to measure the length of the opening of a bay. Using the illustration and various measurements determined on land, find the length of the opening of the bay.

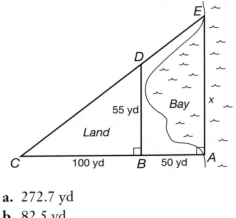

a. 272.7 yd
b. 82.5 yd
c. 27.5 yd
d. 205 yd

429. A car is initially 200 meters due west of a roundabout (traffic circle). If the car travels to the roundabout, continues halfway around the circle, exits due east, then travels an additional 160 meters, what is the total distance the car has traveled? Refer to diagram below. (Use π = 3.14)

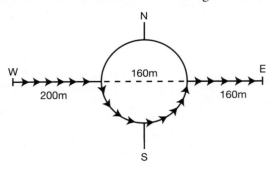

a. 862.4 m
b. 611.2 m
c. 502.4 m
d. 451.2 m

430. Brian Turner is approaching the shores of Australia on his first successful solo hot air balloon ride around the world. His balloon is being escorted by a boat (directly below him) that is 108 meters away. The boat is 144 meters from the shore. How far is the balloon from the shore?

  a. 252 m
  b. 95.2 m
  c. 126 m
  d. 180 m

431. Computer monitors are measured by their diagonals. If a monitor is advertised to be 17 in, what is the actual viewing area, assuming the screen is square?

  a. 289 in²
  b. 90.25 in²
  c. 144.4 in²
  d. 144.5 in²

432. An elevated, cylindrical-shaped water tower is in need of paint. If the radius of the tower is 10 ft and the height is 40 ft, what is the total area to be painted? (Use π = 3.14)

  a. 1,570 ft²
  b. 2,826 ft²
  c. 2,575 ft²
  d. 3,140 ft²

433. A sinking ship signals to the shore for help. Three individuals spot the signal from shore. The first individual is directly perpendicular to the sinking ship and 20 meters inland. The second individual is also 20 meters inland but 100 meters to the right of the first individual. The third is also 20 meters inland but 125 meters to the right of the first individual. How far off shore is the sinking ship? See illustration.

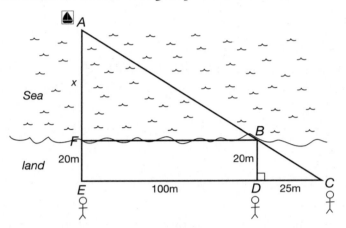

a. 60 meters

b. 136 meters

c. 100 meters

d. 80 meters

434. You are painting the surface of a silo that has a radius of 8 ft and a height of 50 ft. What is the total surface area to be painted? Assume the top of the silo is $\frac{1}{2}$ a sphere and that the silo sits on the ground. Refer to the illustration. (Use $\pi = 3.14$)

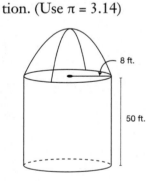

a. 2,913.92 ft²

b. 1,607.68 ft²

c. 2,612.48 ft²

d. 3,315.84 ft²

The Washington Monument is located in Washington, DC. Use the following illustration, which represents one of four identical sides, to answer questions 435 and 436.

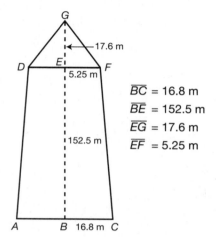

$\overline{BC}$ = 16.8 m
$\overline{BE}$ = 152.5 m
$\overline{EG}$ = 17.6 m
$\overline{EF}$ = 5.25 m

435. Find the height of the Washington Monument to the nearest tenth of a meter.
   a. 157.8 m
   b. 169.3 m
   c. 170.1 m
   d. 192.2 m

436. Find the surface area of the monument to the nearest square meter.
   a. 13,820 m²
   b. 13,451 m²
   c. 3,455 m²
   d. 13,543 m²

437. An inground pool is filling with water. The shallow end is 3 ft deep and gradually slopes to the deepest end, which is 10 ft deep. The width of the pool is 15 ft, and the length is 30 ft. What is the volume of the pool?

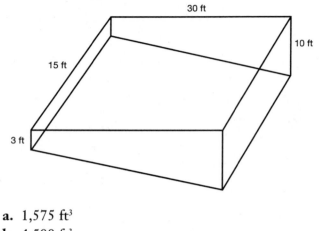

    a. 1,575 ft³

    b. 4,500 ft³

    c. 2,925 ft³

    d. 1,350 ft³

For questions 438 and 439, refer to the following illustration:

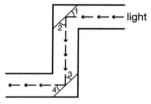

438. In a periscope, two mirrors are mounted parallel to each other as shown. The path of light becomes a transversal. If $\angle 2$ measures 50°, what is the measurement of $\angle 3$?

    a. 50°

    b. 40°

    c. 130°

    d. 310°

439. Given that ∠2 measures 50°, what is the measurement of ∠4?
   a. 50°
   b. 40°
   c. 130°
   d. 85°

440. The angle measure of each of the base angles of an isosceles triangle is represented by x, and the vertex angle is $3x + 20$. Find the measure of a base angle.
   a. 116°
   b. 40°
   c. 32°
   d. 14°

441. Using the information from question 440, find the measure of the vertex angle of the isosceles triangle.
   a. 32°
   b. 62°
   c. 58°
   d. 116°

442. In parallelogram $ABCD$, ∠A = $5x + 2$ and ∠C = $6x - 4$. Find the measure of ∠A.
   a. 32°
   b. 6°
   c. 84.7°
   d. 44°

443. The length of the longer base of a trapezoid is three times the length of the shorter base. The nonparallel sides are congruent. The nonparallel sides are 5 cm longer than the shorter base. The perimeter of the trapezoid is 40 cm. What is the length of the longer base?
   a. 15 cm
   b. 5 cm
   c. 10 cm
   d. 21 cm

444. The measure of the angles of a triangle are represented by $2x + 15$, $x + 20$, and $3x + 25$. Find the measure of the smallest angle.

a. 40°

b. 85°

c. 25°

d. 55°

445. Suppose *ABCD* is a rectangle. IF $\overline{AB}$ = 10 and $\overline{AD}$ = 6, find $\overline{BX}$ to the nearest tenth.

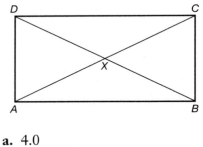

a. 4.0

b. 5.8

c. 11.7

d. 8.0

446. The perimeter of the parallelogram below is 32 cm. What is the length of the longer side?

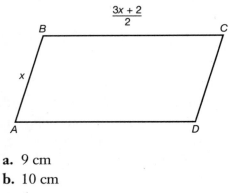

a. 9 cm

b. 10 cm

c. 6 cm

d. 12 cm

447. A door is 6 feet 6 inches tall and 36 inches wide. What is the widest piece of sheetrock that can fit through the door? Round to the nearest inch.
    a. 114 in.
    b. 86 in.
    c. 85 in.
    d. 69 in.

448. The width of a rectangle is 20 cm. The diagonal is 8 cm longer than the length. Find the length of the rectangle.
    a. 20 cm
    b. 23 cm
    c. 22 cm
    d. 21 cm

449. The measures of two complementary angles are in the ratio of 5:7. Find the measure of the smaller angle.
    a. 75°
    b. 37.5°
    c. 52.5°
    d. 105°

450. A pool is surrounded by a deck that is the same width all the way around. The total area of the deck only is 400 square feet. The dimensions of the pool are 18 feet by 24 feet. How many feet wide is the deck?
    a. 4
    b. 8
    c. 24
    d. 25

451. Using the diagram below and the fact that $m\angle A + m\angle B + m\angle C + m\angle D = 320°$, find $m\angle E$.

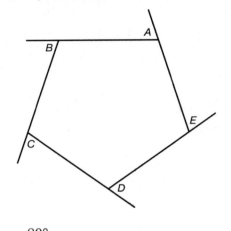

    a. 80°

    b. 40°

    c. 25°

    d. 75°

452. The base of a triangle is 4 times as long as its height. If together they measure 95 cm, what is the area of the triangle?

    a. 1,444 cm²

    b. 100 cm²

    c. 722 cm²

    d. 95 cm2

453. One method of finding the height of an object is to place a mirror on the ground and then position yourself so that the top of the object can be seen in the mirror. How tall is a structure if a person who is 160 cm tall observes the top of a structure when the mirror is 100 m from the structure and the person is 8 m from the mirror?

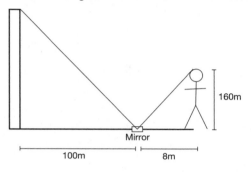

100m   8m   Mirror   160m

   a. 50,000 cm
   b. 20,000 cm
   c. 2,000 cm
   d. 200 cm

454. Suppose *ABCD* is a parallelogram; $m\angle 6 = 120°$ and $m\angle 2 = 40°$. Find $m\angle 4$.

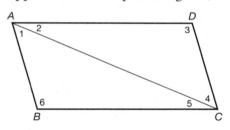

   a. 50°
   b. 40°
   c. 20°
   d. 30°

455. Jessica has a picture in a frame with a total area of 288 in². The dimensions of the picture without the frame are 12 in. by 14 in. What is the larger dimension of the frame, in inches?

    a. 2

    b. 14

    c. 18

    d. 16

456. A sphere has a volume of 288π cm³. Find its radius.

    a. 9.5 cm

    b. 7 cm

    c. 14 cm

    d. 6 cm

457. Using the illustration provided below, if $m\angle ABE = 4x + 5$ and $m\angle CBD = 7x - 13$, find the measure of $\angle ABE$.

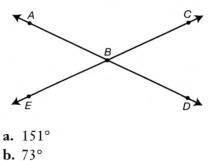

    a. 151°

    b. 73°

    c. 107°

    d. 29°

458. Two angles are complementary. The measure of one angle is four times the measure of the other. Find the measure of the larger angle.

    a. 36°

    b. 72°

    c. 144°

    d. 18°

459. If Gretta's bicycle has a 25-inch diameter wheel, how far will she travel in two turns of the wheel? (Use $\pi = 3.14$)
   a. 491 in.
   b. 78.5 in.
   c. 100 in.
   d. 157 in.

460. Two angles are supplementary. The measure of one angle is 30° more than twice the measure of the other. Find the measure of the larger angle.
   a. 130°
   b. 20°
   c. 50°
   d. 70°

461. Using the illustration provided, find $m\angle AED$, given $m\angle BEC = 5x - 36$ and $m\angle AED = 2x + 9$.

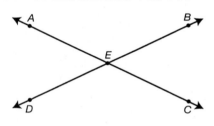

   a. 141°
   b. 69°
   c. 111°
   d. 39°

462. The measures of the angles of a triangle are in the ratio 3:4:5. Find the measure of the largest angle.
   a. 75°
   b. 37.5°
   c. 45°
   d. 60°

463. A mailbox opening is 4.5 inches high and 5 inches wide. What is the widest piece of mail that is able to fit in the mailbox without bending? Round your answer to the nearest tenth of an inch.
   a. 9.5 inches
   b. 2.2 inches
   c. 6.7 inches
   d. 8.9 inches

464. The figure below represents the cross section of a pipe $\frac{1}{2}$ inch thick that has an inside diameter of 3 inches. Find the area of the shaded region in terms of π.

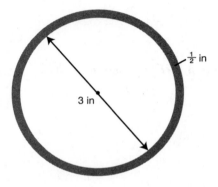

   a. 8.75π in^2
   b. 3.25π in^2
   c. 7π in^2
   d. 1.75π in^2

465. Using the same cross section of pipe from question 464, answer the following question: If the pipe is 18 inches long, what is the volume of the shaded region in terms of π?
   a. 31.5π in^3
   b. 126π in^3
   c. 157.5 in^3
   d. 58.5 in3

466. A person travels 10 miles due north, 4 miles due west, 5 miles due north, and 12 miles due east. How far is this person from the starting point?

    a. 23 miles

    b. 13 miles

    c. 17 miles

    d. 20 miles

467. Using the illustration below, find the area of the shaded region in terms of π.

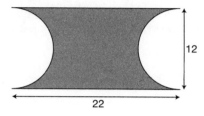

    a. $(264 - 18\pi)$ square units

    b. $(264 - 36\pi)$ square units

    c. $(264 - 12\pi)$ square units

    d. $(18\pi - 264)$ square units

468. Find how many square centimeters of paper are needed to create a label on a cylindrical can 45 cm tall with a circular base having a diameter of 20 cm. Leave your answer in terms of π.

    a. 450π cm^2

    b. $4,500\pi$ cm^2

    c. 900π cm^2

    d. $9,000\pi$ cm^2

469. The height of the unabridged *American Heritage Dictionary* is 2 inches greater than its width. If the perimeter of the dictionary is 48 inches, what is the area of a typical page of the dictionary?

　　a. 48 square inches
　　b. 121 square inches
　　c. 143 square inches
　　d. 169 square inches

470. The structural support system for a bridge is shown in the illustration below. $\overline{AD}$ is parallel to $\overline{BC}$, $\overline{BE}$ is parallel to $\overline{CD}$, and $\overline{AB}$ is parallel to $\overline{CF}$. Find in $m\angle CGE$.

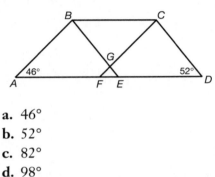

　　a. 46°
　　b. 52°
　　c. 82°
　　d. 98°

471. Find the area of the shaded areas in the figure below, where $\overline{AB}$ = 6 and $\overline{BC}$ = 10. Leave your answer in terms of π.

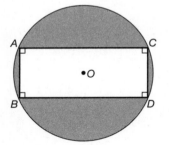

　　a. (25π – 72) square units
　　b. (25π – 48) square units
　　c. (25π – 8) square units
　　d. (100π – 48) square units

472. Find the area of the shaded region in the figure below in terms of π.

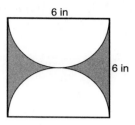

6 in

6 in

    a. (12 – 9π) square units
    b. (36 – 9π) square units
    c. (36 – 4.5π) square units
    d. (2π – 16) square units

473. On a piece of machinery, the centers of two pulleys are 3 feet apart, and the radius of each pulley is 6 inches. How long a belt (in feet) is needed to wrap around both pulleys? Leave your answer in terms of π.

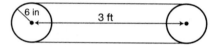

6 in   3 ft

    a. (6 + 0.5π) ft
    b. (6 + 0.25π) ft
    c. (6 + 12π) ft
    d. (6 + π) ft

474. Find the measure of each angle of a regular 15-sided polygon to the nearest tenth of a degree.
    a. 24°
    b. 12°
    c. 128.6°
    d. 156°

475. A sand pile is shaped like a cone as illustrated below. How many cubic yards of sand are in the pile? Round to the nearest tenth of a cubic yard. (Use π = 3.14)

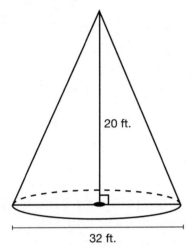

20 ft.

32 ft.

a. 5,358.9 yd³
b. 595.4 yd³
c. 198.5 yd³
d. 793.9 yd3

476. Find the area of the regular octagon with the following measurements.

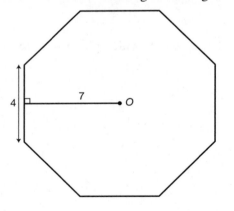

4   7   O

a. 224 square units
b. 112 square units
c. 84 square units
d. 169 square units

477. Two sides of a picture frame are glued together to form a corner. Each side is cut at a 45-degree angle. Using the illustration provided, find the measure of ∠A.

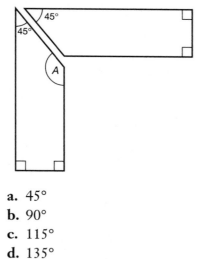

    a. 45°
    b. 90°
    c. 115°
    d. 135°

478. Find the total area of the shaded regions in the illusion below, if the radius of each circle is 5 cm. Leave your answer in terms of π.

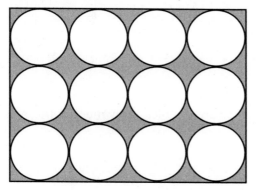

    a. (1,200 – 300π) cm²
    b. (300 – 300π) cm²
    c. (300π – 1,200) cm²
    d. (300π – 300) cm²

479. The road from town A to town B travels at a direction of N23°E. The road from town C to town D travels at a direction of S48°E. The roads intersect at location E. Find the measure of ∠BED, at the point of intersection.

    a. 71°

    b. 23°

    c. 109°

    d. 48°

480. The figure provided below represents a hexagonal-shaped nut. What is the measure of ∠ABC?

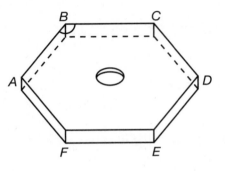

    a. 120°

    b. 135°

    c. 108°

    d. 144°

481. If the lengths of all sides of a box are doubled, how much is the volume increased?

    a. 2 times

    b. 4 times

    c. 6 times

    d. 8 times

482. If the radius of a circle is tripled, the circumference is

    a. multiplied by 3.

    b. multiplied by 6.

    c. multiplied by 9.

    d. multiplied by 12.

483. If the diameter of a sphere is doubled, the surface area is
   a. multiplied by 4.
   b. multiplied by 2.
   c. multiplied by 3.
   d. multiplied by 8.

484. If the diameter of a sphere is tripled, the volume is
   a. multiplied by 3.
   b. multiplied by 27.
   c. multiplied by 9.
   d. multiplied by 6.

485. If the radius of a right circular cone is doubled, the volume is
   a. multiplied by 2.
   b. multiplied by 4.
   c. multiplied by 6.
   d. multiplied by 8.

486. If the radius of a right circular cone is halved, the volume is
   a. multiplied by $\frac{1}{4}$.
   b. multiplied by $\frac{1}{2}$.
   c. multiplied by $\frac{1}{8}$.
   d. multiplied by $\frac{1}{16}$.

487. If the radius of a right cylinder is doubled and the height is halved, its volume
   a. remains the same.
   b. is multiplied by 2.
   c. is multiplied by 4.
   d. is multiplied by $\frac{1}{2}$.

488. If the radius of a right cylinder is doubled and the height is tripled, its volume is
   a. multiplied by 12.
   b. multiplied by 2.
   c. multiplied by 6
   d. multiplied by 3.

489. If each interior angle of a regular polygon has a measure of 150 degrees, how many sides does the polygon have?

    a. 10

    b. 11

    c. 12

    d. 13

490. A box is 30 cm long, 8 cm wide, and 12 cm high. How long is the diagonal $\overline{AB}$? Round to the nearest tenth of a centimeter.

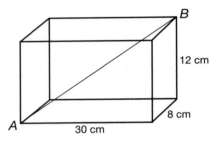

    a. 34.5 cm

    b. 32.1 cm

    c. 35.2 cm

    d. 33.3 cm

491. Find the area of the shaded region in the figure below. Leave your answer in terms of π.

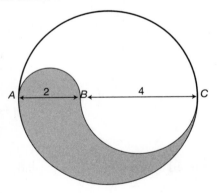

    a. 16.5π square units

    b. 30π square units

    c. 3π square units

    d. 7.5π square units

492. A round tower with a 40 meter circumference is surrounded by a security fence that is 8 meters from the tower. How long is the security fence in terms of π?

  a. (40 + 16π) meters
  b. (40 + 8π) meters
  c. 48π meters
  d. 56π meters

493. The figure below consists of two overlapping rectangles. Find the sum of $m\angle 1 + m\angle 2 + m\angle 3 + m\angle 4$.

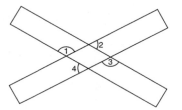

  a. 360°
  b. 90°
  c. 180°
  d. 540°

494. A solid is formed by cutting the top off of a cone with a slice parallel to the base, and then cutting a cylindrical hole into the resulting solid. Find the volume of the hollow solid in terms of π.

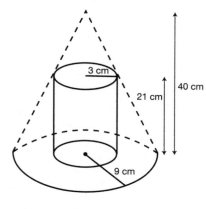

  a. 834π cm³
  b. 2,880π cm³
  c. 891π cm³
  d. 1,326π cm3

495. A rectangular container is 5 cm wide and 15 cm long, and contains water to a depth of 8 cm. An object is placed in the water, and the water rises 2.3 cm. What is the volume of the object?

a. 92 cm³

b. 276 cm³

c. 172.5 cm³

d. 312.5 cm3

496. A concrete retaining wall is 120 feet long with ends shaped as shown. How many cubic yards of concrete are needed to construct the wall?

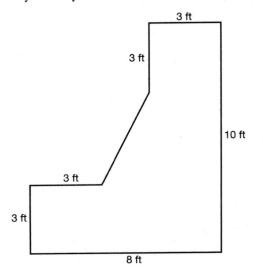

a. 217.8 yd³

b. 5,880 yd³

c. 653.3 yd³

d. 49 yd³

497. A spherical holding tank whose diameter to the outer surface is 20 feet is constructed of steel 1 inch thick. How many cubic feet of steel is needed to construct the holding tank? Round to the nearest cubic foot. (Use π = 3.14)

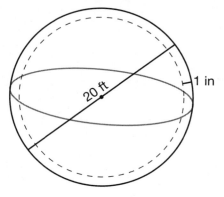

   a. 78 ft³
   b. 104 ft³
   c. 26 ft³
   d. 125 ft³

498. How many cubic inches of lead are there in the pencil? Round to the nearest thousandth of a cubic inch. (Use π = 3.14)

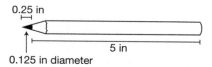

   a. 0.061 in³
   b. 0.060 in³
   c. 0.062 in³
   d. 0.063 in³

499. A cylindrical hole with a diameter of 4 inches is cut through a cube. The edges of the cube measure 5 inches. Find the volume of the hollow solid in terms of π.

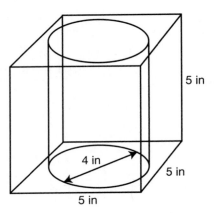

5 in

4 in

5 in

5 in

a. $(125 - 80\pi)$ in^3
b. $(125 - 20\pi)$ in^3
c. $(80\pi - 125)$ in^3
d. $(20\pi - 125)$ in^3

500. Find the area of the region illustrated below.

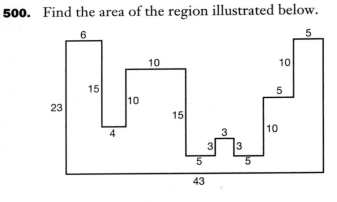

a. 478 units^2
b. 578 units^2
c. 528 units^2
d. 428 units2

501. From a stationary point directly in front of the center of a bull's eye, Kim aims two arrows at the bull's eye. The first arrow nicks one point on the edge of the bull's eye; the second strikes the center of the bull's eye. Kim knows the second arrow traveled 20 meters since she knows how far she is from the target. If the bull's eye is 4 meters wide, how far did the first arrow travel? You may assume that the arrows traveled in straight-line paths and that the bull's eye is circular. Round your answer to the nearest tenth of a meter.

   a. 19.9 meters
   b. 24 meters
   c. 22 meters
   d. 20.1 meters

Answer Explanations

The following explanations show one way in which each problem can be solved. You may have another method for solving these problems.

376. **d.** The area of a rectangle the is *length* times the *width*.

377. **b.** The volume of a sphere is $\frac{4}{3}$ times π times the radius cubed.

378. **b.** The area of a triangle is $\frac{1}{2}$ times the length of the base times the length of the height.

379. **b.** The surface area of a sphere is four times π times the radius squared.

380. **d.** The area of a circle is π times the radius squared.

381. **a.** The volume of a cylinder is π times the radius squared, times the height of the cylinder.

382. **c.** The perimeter of a square is four times the length of one side.

383. **c.** The area of the base is π times the radius squared. The area of the curved region is two times π times the radius times the height. Notice there is only one circular region since the storage tank would be on the ground. This area would not be painted.

384. **a.** The area of a square is the length of one side squared (or side times side).

385. **c.** The circumference, or distance around a circle, is π times the diameter.

386. **b.** Let x = the number of inches in the width and let $x + 5$ = the number of inches in the length. Since the area of a rectangle is *length* $\times$ *width*, the equation for the area of the rectangle is $x(x + 5) = 24$. Multiply the left side of the equation using the distributive property: $x^2 + 5x = 24$. Put the equation in the standard form $ax^2 + bx + c = 0$: $x^2 + 5x - 24 = 0$. Factor the left side of the equation: $(x + 8)(x - 3) = 0$. Set each factor equal to zero and solve: $x + 8 = 0$ or $x - 3 = 0$; $x = {}^-8$ or $x = 3$. Reject the solution of $^-8$ because a the width of a rectangle cannot be negative. The width is 3 inches.

387. d. The volume of a cube is the length of one side cubed (or the length of the side times the length of the side times the length of the side).

388. a. The perimeter of a triangle is the length of side *a* plus the length of side *b* plus the length of side *c*.

389. c. The area of a rectangle is *length × width*. Therefore, the area of the racquetball court is equal to 40 ft times 20 ft, or 800 ft². If you chose answer **d**, you found the perimeter of, or distance around, the court.

390. a. The width of the field, 180 ft, must be divided by the width of the mower, 2 ft. The result is that he must mow across the lawn 90 times. If you chose answer **b**, you calculated as if he were mowing the length of the field. If you chose answer **c**, you combined length and width, which would result in mowing the field twice.

391. a. The area to be painted is the sum of the areas of the four rectangular walls. Each wall has an area of *length × width*, or (9)(5.5), which equals 49.5 ft². Multiply this by 4 which equals 198 ft². If you chose **b**, you added 9 ft and 5.5 ft instead of multiplying.

392. b. Let x = the measure of the base and $x + 5$ = the measure of the height. Since the area of a parallelogram is *base × height*, then the equation for the area of the parallelogram is $x(x + 5) = 36$. Multiply the left side of the equation using the distributive property: $x^2 + 5x = 36$; Put the equation in standard form and set it equal to zero: $x^2 + 5x - 36. = 0$. Factor the left side of the equation: $(x + 9)(x - 4) = 0$. Set each factor equal to zero and solve: $x + 9 = 0$ or $x - 4 = 0$; $x = {}^-9$ or $x = 4$. Reject the solution of $^-9$ because the length of a parallelgram's base cannot be negative. The height is $4 + 5 = 9$ meters.

393. c. To find the height of Heather's poodle, set up a proportion with $\frac{building\ height}{building\ shadow} = \frac{poodle\ height}{poodle\ shadow}$: $\frac{45}{30} = \frac{x}{2.5}$. Cross-multiply: $112.5 = 30x$. Solve for x: 3.75 ft $= x$. If you chose **d**, the proportion was set up incorrectly as $\frac{45}{30} = \frac{2.5}{x}$.

394. b. The volume of a cylinder is $\pi r^2 h$. Using a height of 4 ft and radius of 10 ft, the volume of the pool is $(3.14)(10)^2(4)$ or 1,256 ft^3. If you chose **a**, you used πdh instead of $\pi r^2 h$. If you chose **c**, you used the diameter squared instead of the radius squared.

395. b. The connection of the pole with the ground forms the right angle of a triangle. The length of the pole is a leg of the right triangle. The distance between the stake and the pole is also a leg of the right triangle. The question is to find the length of the cable, which is the hypotenuse. Using the Pythagorean theorem: $(36)^2 + (15)^2 = c^2$; $1,296 + 225 = c^2$; $1,521 = c^2$; 39 ft $= c$.

396. b. The distance around the room is $2(12) + 2(13) = 50$ ft. Each roll of border is $5(3)$ or 15 ft long. By dividing the total distance around the room, 50 ft, by the length of each roll, 15 ft, we find we need 3.33 rolls. Since a roll cannot be subdivided, 4 rolls will be needed.

397. d. Let x = the length of the diagonal. Therefore, $x - 5$ = the length of the patio and $x - 7$ = the width of the patio. Since the area is 195 m^2, and *Area = length × width*, the equation is $(x - 5)(x - 7) = 195$. Use the distributive property to multiply the binomials: $x^2 - 5x - 7x + 35 = 195$. Combine like terms: $x^2 - 12x + 35 = 195$. Subtract 195 from both sides: $x^2 - 12x + 35 - 195 = 195 - 195$. Simplify: $x^2 - 12x - 160 = 0$. Factor the result: $(x - 20)(x + 8) = 0$. Set each factor equal to 0 and solve: $x - 20 = 0$ or $x + 8 = 0$; $x = 20$ or $x = -8$. Reject the solution of -8 because the length of a diagonal cannot be negative. The length of the diagonal is 20 meters.

398. a. Let w = the width of the field and $2w + 2$ = the length of the field (two feet more than twice the width). Since *Area = length × width*, multiply the two expressions together and set them equal to 3,280: $w(2w + 2) = 3,280$. Multiply using the distributive property: $2w^2 + 2w = 3,280$. Subtract 3,280 from both sides: $2w^2 + 2w - 3,280 = 3,280 - 3,280$; simplify: $2w^2 + 2w - 3,280 = 0$. Factor the trinomial completely: $2(w^2 + w - 1640) = 0$; $2(w + 41)(w - 40) = 0$. Set each factor equal to zero and solve: $2 \neq 0$ or $w + 41 = 0$ or $w - 40 = 0$; $w = -41$ or $w = 40$. Reject the negative solution because the width of a field cannot be negative. The width is 40 feet.

399. b. To find how far Shannon will need to travel, set up the following proportion: $\frac{1 \text{ inch}}{14 \text{ miles}} = \frac{17 \text{ inches}}{x \text{ miles}}$. Cross multiply: $x = (17)(14)$; $x = 238$ miles.

400. a. The circumference of a circle is πd. Using the diameter of 10 inches, the circumference is equal to $(3.14)(10)$, or 31.4 inches. If you chose **b**, you found the area of a circle. If you chose **c**, you mistakenly used πr for circumference rather than $2\pi r$. If you chose **d**, you used $2\pi d$ rather than $2\pi r$.

401. c. The circumference of a circle is πd. Since 10 ft represents the radius, the diameter is 20 feet, because the diameter of a circle is twice the radius. Therefore, the circumference is $(3.14)(20)$, or 62.8 ft. If you chose **a**, you used πr rather than $2\pi r$. If you chose **b**, you found the area rather than circumference.

402. a. The area of a triangle is $\frac{1}{2}(base)(height)$. Using the dimensions given, $Area = \frac{1}{2}(40)(83) = 1,660$ ft². If you chose **d**, you omitted $\frac{1}{2}$ from the formula.

403. b. The volume of a sphere is $\frac{4}{3}\pi r^3$. Using the dimensions given, $Volume = \frac{4}{3}(3.14)(4)^3 = 267.9$. Rounding this answer to the nearest cubic inch is 268 in³. If you chose **a**, you found the surface area rather than volume. If you chose **c**, you miscalculated the volume by using the diameter.

404. c. The area of a circle is πr^2. The diameter = 42 in, so the radius = 42 ÷ 2 = 21 in; $(3.14)(21)^2 = 1,384.74$ in². Rounding to the nearest square inch, the answer is 1,385 in². If you chose **a**, you rounded the final answer incorrectly. If you chose **d**, you used the diameter instead of the radius.

405. d. To find the volume of a sphere, use the formula $Volume = \frac{4}{3}\pi r^3$. $Volume = \frac{4}{3}(3.14)(3.25)^3 = 143.72$ in³. If you chose **a**, you squared the radius instead of cubing it. If you chose **b**, you cubed the diameter instead of the radius. If you chose **c**, you found the surface area of the sphere, not the volume.

406. c. The ladder forms a right triangle with the building. The length of the ladder is the hypotenuse, and the distance of the ladder from the base of the building is a leg. The question asks you to solve for the remaining leg of the triangle, or how far up the building the ladder will reach. Use the Pythagorean theorem: $9^2 + b^2 = 15^2$; $81 + b^2 = 225$; $81 + b^2 - 81 = 225 - 81$; $b^2 = 144$; $b = 12$ ft.

407. b. The volume of a rectangular solid is *length × width × height*. Using the dimensions in the question, Volume = $(22)(5)(5) = 550$ in^3. If you chose **c**, you found the surface area of the box.

408. b. A minute hand moves 180 degrees in 30 minutes. Use the following proportion: $\frac{30 \text{ minutes}}{180 \text{ degrees}} = \frac{25 \text{ minutes}}{x \text{ degrees}}$. Cross-multiply: $30x = 4{,}500$. Solve for x: $x = 150$ degrees.

409. a. The planes are traveling perpendicular to each other. The courses they are traveling form the legs of a right triangle. The question requires us to find the distance between the planes, or the length of the hypotenuse. Use the Pythagorean theorem: $70^2 + 168^2 = c^2$; $4{,}900 + 28{,}224 = c^2$; $33{,}124 = c^2$; $c = 182$ miles. If you chose **c**, you assigned the hypotenuse the value of 168 miles and solved for a leg rather than the hypotenuse. If you chose **d**, you added the legs together rather than using the Pythagorean theorem.

410. d. The area of a small pizza is 78.5 in^2. The question requires us to find the diameter of the pizza in order to select the most appropriate box. Area is equal to πr^2. Therefore, $78.5 = \pi r^2$; divide by π (3.14); $78.5 \div 3.14 = \pi r^2 \div 3.14$; $25 = r^2$; $5 = r$. The diameter is twice the radius or 10 inches. Therefore, the box is also 10 inches.

411. d. The area of Stuckeyburg can be found by dividing the region into a rectangle and a triangle. Find the area of the rectangle ($A = lw$) and add the area of the triangle ($\frac{1}{2}bh$) for the total area of the region. Referring to the diagram, the area of the rectangle is $(10)(8) = 80$ miles^2. The area of the triangle is $\frac{1}{2}(8)(3) = 12$ miles^2. The sum of the two regions is 80 miles^2 + 12 miles^2 = 92 miles^2. If you chose **a**, you found the perimeter of the town. If you chose **b**, you found the area of the rectangular region but did not include the triangular region.

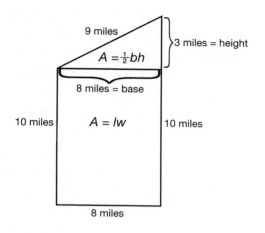

9 miles

3 miles = height

$A = \frac{1}{2}bh$

8 miles = base

10 miles $A = lw$ 10 miles

8 miles

412. b. The area of a rectangle is *length* × *width*. Using the formula 1,960 yd² = (*l*)(28), solve for *l* by dividing both sides by 28: *l* = 70 yards.

413. b. To find the area of the matting, subtract the area of the print from the area of the frame. The area of the print is found using πr^2 or $(3.14)(7)^2$, which equals 153.86 in². The area of the square frame is s^2 or (18)(18), which equals 324 in². The difference, 324 in² – 153.86 in² or 170.14 in², is the area of the matting. If you chose **c**, you mistakenly used the formula for the circumference of a circle, $2\pi r$, instead of the area of a circle, πr^2.

414. a. The ribbon will travel the length (10 in) twice, the width (7 in) twice, and the height (4 in) four times. Adding up these distances will determine the total amount of ribbon needed: 10 in + 10 in + 7 in + 7 in + 4 in + 4 in + 4 in + 4 in = 50 inches of ribbon. If you chose **b**, you missed two sides of 4 inches. If you chose **d**, you calculated the volume of the box.

415. d. To find the area of the skirt, find the area of the outer circle minus the area of the inner circle. The area of the outer circle is $\pi(3.5)^2$ or 38.465 in². The area of the inner circle is $\pi(.5)^2$ or .785 in². The difference is 38.465 – 0.785 or 37.68 ft². The answer, rounded to the nearest foot, is 38 ft². If you chose **a**, you rounded to the nearest tenth of a foot. If you chose **b**, you miscalculated the radius of the outer circle as being 3 feet instead of 3.5 feet.

416. **c.** Since the tiles are measured in inches, convert the area of the floor to inches as well. The length of the floor is 9 ft × 12 inches per foot = 108 in. The width of the floor is 11 ft × 12 inches per foot = 132 in. The formula for the area of a rectangle is *length × width*. Therefore, the area of the kitchen floor is 108 in × 132 in = 14,256 in². The area of one tile is 6 in × 6 in, or 36 in². Finally, divide the total number of square inches by 36 in²: 14,256 in² ÷ 36 in²: per tile = 396 tiles.

417. **a.** If a framed print is enclosed by a 2-inch matting, the print is 4 inches less in length and height. Therefore, the picture is 32 in by 18 in. These measurements along with the diagonal form a right triangle. Using the Pythagorean theorem, solve for the diagonal: $32^2 + 18^2 = c^2$; $1,024 + 324 = c^2$; $1,348 = c^2$; $36.7 = c$. If you chose **b**, you reduced the print 2 inches less than the frame in length and height rather than 4 inches.

418. **a.** Let $x =$ the width of the walkway. Since the width of the garden only is 24, the width of the garden and the walkway together is $x + x + 24$, or $2x + 24$. Since the length of the garden only is 35, the length of the garden and the walkway together is $x + x + 35$, or $2x + 35$. The area of a rectangle is *length × width*, so multiply the expressions together and set the result equal to the total area of 1,530 square feet: $(2x + 24)(2x + 35) = 1,530$. Multiply the binomials using the distributive property: $4x2 + 70x + 48x + 840 = 1,530$. Combine like terms: $4x^2 + 118x + 840 = 1,530$. Subtract 1,530 from both sides: $4x^2 + 118x + 840 - 1,530 = 1,530 - 1,530$; simplify: $4x^2 + 118x - 690 = 0$. Factor the trinomial completely: $2(2x^2 + 59x - 345) = 0$; $2(2x + 69)(x - 5) = 0$; $x = -34.5$ or $x = 5$. Reject the negative solution because the width of a walkway cannot be negative. The width is 5 feet.

419. **c.** The surface area of the box is the sum of the areas of all six sides. Two sides are 20 in by 18 in, or $(20)(18) = 360$ in². Two sides are 18 in by 4 in, or $(18)(4) = 72$ in². The last two sides are 20 in by 4 in, or $(20)(4) = 80$ in². Add up all six sides to get the total area: 360 in² + 360 in² + 72 in² + 72 in² + 80 in² + 80 in² = 1,024 in². If you chose **a**, you did not double all sides. If you chose **b**, you calculated the volume of the box.

420. **d.** The area of the walkway is equal to the entire area (area of the walkway and pool) minus the area of the pool. The area of the entire rectangular region is *length × width*. Since the pool is 20 feet wide and the walkway adds 4 feet onto each side, the width of the rectangle formed by the walkway and the pool is 20 + 4 + 4 = 28 feet. Since the pool is 40 feet long and the walkway adds 4 feet onto each side, the length of the rectangle formed by the walkway and the pool is 40 + 4 + 4 = 48 feet. Therefore, the area of the walkway and the pool is (28)(48) = 1,344 ft². The area of the pool is (20)(40) = 800 ft², so the area of just the walkway is 1,344 ft² – 800 ft² = 544 ft². If you chose **c**, you extended the entire area's length and width by 4 feet instead of 8 feet.

421. b. The area described as section A is a trapezoid. The formula for the area of a trapezoid is $\frac{1}{2}h(b_1 + b_2)$. The height of the trapezoid is 2 inches, b_1 is 6 inches, and b_2 is 8 inches. Using these dimensions, area = $\frac{1}{2}(2)(6 + 8)$ or 14 in². If you chose **a**, you found the area of section B or D.

422. b. To find the total area, add the area of region A plus the area of region B plus the area of region C. The area of region A is *length × width* or (100)(40) = 4,000 m². Area of region B is $\frac{1}{2}bh$ or $\frac{1}{2}(40)(30)$ = 600 m². The area of region C is $\frac{1}{2}bh$ or $\frac{1}{2}(30)(40)$ = 600 m². The combined area is the sum of the three areas: 4,000 + 600 + 600 = 5,200 m². If you chose **a**, you miscalculated the area of a triangle as bh instead of $\frac{1}{2}bh$. If you chose **c**, you found only the area of the rectangle. If you chose **d**, you found the area of the rectangle and only one of the triangles.

423. c. To find the perimeter of the plot of land, we must know the length of all sides. According to the diagram, we must find the length of the hypotenuse for the triangular regions B and C. Using the Pythagorean theorem for triangular region B, $30^2 + 40^2 = c^2$; 900 + 1,600 = c^2; 2,500 = c^2; 50 m = c. The hypotenuse for triangular region C is also 50 m since the legs are 30 m and 40 m as well. Now add the length of all sides: 40 m + 100 m + 30 m + 50 m + 30 m + 50 m + 60 m = 360 m, which is the perimeter of the plot of land. If you chose **a**, you did not calculate in the hypotenuse on either triangle. If you chose **b**, you miscalculated the hypotenuses as having lengths of 40 m. If you chose **d**, you miscalculated the hypotenuses as having lengths of 30 m.

424. d. The 18 ft pole is perpendicular to the ground, forming the right angle of a triangle. The 22 ft guy wire represents the hypotenuse. The task is to find the length of the remaining leg in the triangle. Use the Pythagorean theorem: $18^2 + b^2 = 22^2$; 324 + b^2 = 484; b^2 = 160; $b = \sqrt{160}$ or $4\sqrt{10}$. If you chose **a**, you did not take the square root.

425. c. $\triangle ABD$ is similar to $\triangle ECD$. Using this fact, the following proportion is true: $\frac{DE}{EC} = \frac{DA}{AB}$ or $\frac{40}{32} = \frac{(40 + x)}{60}$. Cross-multiply: $2{,}400 = 32(40 + x)$; $2{,}400 = 1{,}280 + 32x$. Subtract $1{,}280$: $1{,}120 = 32x$; divide by 32: $x = 35$ feet.

426. a. The area of the front cover is *length* $\times$ *width* or $(8)(11) = 88$ in^2. The rear cover has the same area as the front, 88 in^2. The area of the binding is *length* $\times$ *width* or $(1.5)(11) = 16.5$ in^2. The extension inside the front cover is *length* $\times$ *width* or $(2)(11) = 22$ in^2. The extension inside the rear cover is also 22 in^2. The total area of the book jacket is the sum of all previous areas: 88 in^2 + 88 in^2 + 16.5 in^2 + 22 in^2 + 22 in^2 = 236.5 in^2. If you chose **b**, you did not include the area of the extensions inside the front and rear covers. If you chose **c**, you miscalculated the area of the binding as $(1.5)(8)$ and omitted the extensions inside the front and rear covers. If you chose **d**, you miscalculated the area of the binding as $(1.5)(8)$ only.

427. a. To find the area of the rectangular region, multiply *length* $\times$ *width* or $(30)(70)$, which equals $2{,}100$ in^2. To find the area of the semi-circle, multiply $\frac{1}{2}$ times πr^2 or $\frac{1}{2}\pi(15)^2$, which equals 353.25 in^2. Add the two areas together for the area of the entire window: $2{,}100 + 353.25 = 2{,}453.3$, rounded to the nearest tenth. If you chose **b**, you included the area of an entire circle, not a semi-circle.

428. b. νACE and νBCD are similar triangles. Using this fact, the following proportion is true: $\frac{CB}{BD} = \frac{CA}{AE}$ or $\frac{100}{55} = \frac{150}{x}$. Cross-multiply: $100x = 8{,}250$. Divide by 100 to solve for x: $x = 82.5$ yards. If you chose **a** or **c**, you set up the proportion incorrectly.

429. b. The question requires us to find the distance around the semi-circle. This distance will then be added to the distance traveled before entering the roundabout, and the distance traveled after exiting the roundabout. According to the diagram, the diameter of the roundabout is 160 m. The distance or circumference of half a circle is $\frac{1}{2}\pi d$, $\frac{1}{2}(3.14)(160)$ or 251.2 m. The total distance the car travels is 200 m + 160 m + 251.2 m = 611.2 m. If you chose **a**, you included the distance around the entire circle. If you chose **c**, you found only the distance around the circle. If you chose **d**, you did not include the distance traveled after exiting the circle, 160 m.

430. d. The boat is located at the triangle's right angle. The distance between the balloon and the boat is 108 meters, which is one leg of the triangle. The distance between the boat and the land is 144 meters, which is the second leg. The distance between the balloon and the land, which is what we must find, is the hypotenuse. Use the Pythagorean theorem: $108^2 + 144^2 = c^2$; $11,664 + 20,736 = c^2$; $32,400 = c^2$; $c = 180$ m.

431. d. Since the monitor is square, the diagonal and length of the sides of the monitor form an isosceles right triangle. The question requires us to find the length of one leg of the triangle in order to find the area. Use the Pythagorean theorem: $s^2 + s^2 = 17^2$; $2s^2 = 289$. Divide by 2: $s^2 = 144.5$. The area of the square monitor is s^2; thus the actual viewing area of the screen is 144.5 in². If you chose **a**, you simply squared the diagonal, which does not give you the viewing area of the monitor.

432. d. To find the surface area of a cylinder, use the following formula: *Surface Area* $= 2\pi r^2 + \pi dh$. Therefore, the surface area $= 2(3.14)(10)^2 + (3.14)(20)(40) = 3,140$ ft². If, you chose **b**, you found the surface area of the circular top and forgot about the bottom of the water tower. However, the bottom of the tower would need painting since the tank is elevated.

433. d. Using the concept of similar triangles, $\triangle CDB$ is similar to $\triangle CEA$, so set up the following proportion: $\frac{25}{20} = \frac{125}{(x + 20)}$. Cross-multiply: $25x + 500 = 2,500$. Subtract 500: $25x = 2,000$; Divide by 25: $x = 80$ meters. If you chose **b**, the proportion was set up incorrectly as $\frac{25}{20} = \frac{(x + 20)}{125}$.

434. a. To find the total surface area of the silo, add the surface area of the height of the cylinder to the surface area of $\frac{1}{2}$ of the sphere. To find the surface area of the height of the cylinder, use the formula πhd or $(3.14)(50)(16)$, which equals 2,512 ft². The surface area of half a sphere is $(\frac{1}{2})(4)\pi r^2$. Using a radius of 8 ft, the surface area is $(\frac{1}{2})(4)(3.14)(8)^2 = 401.92$ ft². Adding the surface area of the cylinder plus half that of the sphere is $2,512 + 401.92 = 2,913.92$ ft². If you chose **b**, you used the diameter rather than the radius when finding the surface area of $\frac{1}{2}$ the sphere. If you chose **d**, you found the surface area of the entire sphere, not just half.

435. c. The height of the monument is the sum of $\overline{BE}$ and $\overline{EG}$. Therefore, the height is 152.5 m + 17.6 m = 170.1 meters. If you chose **a**, you added $\overline{BE} + \overline{EF}$. If you chose **b**, you added $\overline{BE} + \overline{BC}$.

436. **a.** The surface area of the monument is the sum of 4 sides of a trapezoidal shape plus 4 sides of a triangular shape. The trapezoid $DFCA$ has a height of 152.5 m ($\overline{BE}$), $b_1 = 33.6$ m ($\overline{AC}$), and $b_2 = 10.5$ m ($\overline{DF}$). The area is $\frac{1}{2}h(b_1 + b_2)$ or $\frac{1}{2}(152.5)(33.6 + 10.5)$, which equals 3,362.625 m². The triangle DGF has $b = 10.5$ and $h = 17.6$. The area is $\frac{1}{2}bh$ or $\frac{1}{2}$ (10.5)(17.6), which equals 92.4 m². The sum of 4 trapezoidal regions, (4)(3,362.625) = 13,450.5 m², plus 4 triangular regions, 4(92.4) = 369.6 m², is 13,820.1 m². Rounding this answer to the nearest square meter is 13,820 m². If you chose **b**, you found the area of the trapezoidal regions only. If you chose **c**, you found the area of one trapezoidal region and one triangular region. If you chose **d**, you found the area of 4 trapezoidal regions and one triangular region.

437. **c.** The volume of a rectangular solid is *length × width × height*. First, calculate what the volume would be if the entire pool had a depth of 10 ft. The volume would be (10)(30)(15) or 4,500 ft³. Now subtract the area under the sloped plane, which is a triangular solid, as shown below. The volume of the region is $\frac{1}{2}$(*base*)(*height*)(*depth*) or $\frac{1}{2}$(7)(30)(15) or 1,575 ft³. Subtract to get the volume of the pool: 4,500 ft³ – 1,575 ft³ 2,925 ft³. If you chose **a**, this is the volume of the triangular solid under the sloped plane in the pool. If you chose **b**, you did not calculate the slope of the pool, but rather a pool that is uniformly 10 feet deep.

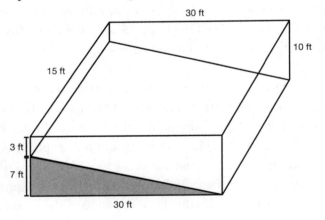

438. a. Two parallel lines cut by a transversal form alternate interior angles that are congruent. The two parallel lines are formed by the mirrors, and the path of light is the transversal. Therefore, $\angle 2$ and $\angle 3$ are alternate interior angles and are congruent. If $\angle 2$ measures 50°, $\angle 3$ is also 50°. If you chose **b**, your mistake was assuming $\angle 2$ and $\angle 3$ are complementary angles. If you chose **c**, your mistake was assuming $\angle 2$ and $\angle 3$ are supplementary angles.

439. b. Knowing that $\angle 4 + \angle 3$ + the right angle placed between $\angle 4$ and $\angle 3$ equals 180, and given the fact that $\angle 3 = 50$, we simply subtract 180 – 90 – 50, which equals 40. If you chose **a**, you assumed that $\angle 3$ and $\angle 4$ are vertical angles. If you chose **c**, you assumed that $\angle 3$ and $\angle 4$ are supplementary.

440. c. The sum of the measures of the angles of a triangle is 180°. The question is asking us to solve for x. The equation is $x + x + 3x + 20 = 180$. Simplifying the equation: $5x + 20 = 180$. Subtract 20 from each side: $5x = 160$. Divide each side by 5: $x = 32°$. If you chose **a**, you solved for the vertex angle. If you chose **b**, you wrote the original equation incorrectly as $x + 3x + 20 = 180$. If you chose **d**, you wrote the original equation incorrectly as $x + x + 3x + 10 = 90$.

441. d. Since we solved for x in the previous question, simply substitute $x = 32°$ into the equation for the vertex angle, $3x + 20$. The result is 116°. If you chose **a**, you solved for the base angle. If you chose **b**, the original equation was written incorrectly as $x + x + 3x + 20 = 90$.

442. a. Opposite angles of a parallelogram are equal in measure. Using this fact, $\angle A = \angle C$, or $5x + 2 = 6x - 4$. Subtract $5x$ from both sides: $2 = x - 4$. Add 4 to both sides: $6 = x$. Now substitute $x = 6$ into the expression for $\angle A$: $5(6) + 2 = 30 + 2$ or $32°$: If you chose **b**, you solved for x, not $\angle A$. If you chose **c**, you assumed the angles were supplementary. If you chose **d**, you assumed the angles were complementary.

443. a. The two bases of the trapezoid are represented by x and $3x$. The non-parallel sides are each $x + 5$. Setting up the equation for the perimeter will allow us to solve for x: $x + 3x + (x + 5) + (x + 5) = 40$. Simplify to $6x + 10 = 40$. Subtract 10 from both sides: $6x = 30$. Divide both sides by 6: $x = 5$ cm. The longer base is represented by $3x$. Using substitution, $3x = (3)(5) = 15$ cm. If you chose **b**, you solved for the shorter base. If you chose **c**, you solved for the nonparallel side. If you chose **d**, the original equation was incorrect.

444. a. The sum of the measures of the angles of a triangle is 180°. Using this information, we can write the equation $(2x + 15) + (x + 20) + (3x + 25) = 180$. Simplify the equation: $6x + 60 = 180$. Subtract 60 from both sides: $6x = 120$. Divide both sides by 6: $x = 20$. Now substitute 20 in for x in each expression to find the smallest angle. The smallest angle is found using the expression $x + 20$: $20 + 20 = 40°$. If you chose **b**, this is the largest angle within the triangle. If you chose **c**, the original equation was incorrectly written as $(2x + 15) + (x + 20) + (3x + 25) = 90$. If you chose **d**, this was the angle that lies numerically between the smallest and largest angle measurements.

445. b. $\overline{AB}$ and $\overline{AD}$ are the legs of a right triangle. $\overline{DB}$ is the hypotenuse, and $\overline{BX}$ is equal to $\frac{1}{2}\overline{DB}$. Solving for the hypotenuse, we use the Pythagorean theorem, $a^2 + b^2 = c^2$: $10^2 + 6^2 = \overline{DB}^2$; $100 + 36 = \overline{DB}^2$; $136 = \overline{DB}^2$. $\overline{DB} = 11.66$; $\frac{1}{2}\overline{DB} = 5.8$. If you chose **a**, you assigned 10 as the length of the hypotenuse. If you chose **d**, the initial error was the same as choice **a**. In addition, you solved for $\overline{DB}$ and not $\frac{1}{2}\overline{DB}$.

446. b. The perimeter of a parallelogram is the sum of the lengths of all four sides. Using this information and the fact that opposite sides of a parallelogram are equal, we can write the following equation: $x + x + \frac{3x + 2}{2} + \frac{3x + 2}{2} = 32$. Simplify to $2x + 3x + 2 = 32$. Simplify again: $5x + 2 = 32$. Subtract 2 from both sides: $5x = 30$. Divide both sides by 5: $x = 6$ cm. The longer base is represented by $\frac{(3x + 2)}{2}$. Using substitution, $\frac{3(6) + 2)}{2} = 10$ cm. If you chose **c**, you solved for the shorter side.

447. b. To find the width of the widest piece of sheetrock that can fit through the door, we recognize it to be equal to the length of the diagonal of the door frame. If the height of the door is 6 ft 6 in, this is equivalent to 78 inches. Using the Pythagorean theorem, we will solve for c: $(78)^2 + (36)^2 = c^2$. Simplify: $6,084 + 1,296 = c^2$; $7,380 = c^2$. Take the square root of both sides: $c = 86$ in. If you chose **a**, you added $78 + 36$. If you chose **c**, you rounded incorrectly. If you chose **d**, you assigned 78 inches as the hypotenuse, c.

448. d. To find the length of the rectangle, we will use the Pythagorean theorem. The width, a, is 20. The diagonal, c, is $x + 8$. The length, b, is x: $a^2 + b^2 = c^2$; $20^2 + x^2 = (x + 8)^2$. After multiplying the two binomials (using FOIL), $400 + x^2 = x^2 + 16x + 64$. Subtract x^2 from both sides: $400 = 16x + 64$. Subtract 64 from both sides: $336 = 16x$. Divide both sides by 16: 21 cm $= x$. If you chose **a**, you incorrectly determined the length of the diagonal to be 28 cm.

449. b. Two angles are complementary if their sum is 90°. Using this fact yields the following equation: $5x + 7x = 90$. Simplify: $12x = 90$. Divide both sides of the equation by 12: $x = 7.5$. The smaller angle is represented by $5x$. Therefore, $5x = 5(7.5)$ or 37.5°, the measurement of the smaller angle. If you chose **a**, the original equation was set equal to 180° rather than 90°. If you chose **c**, you solved for the larger angle. If you chose **d**, the original equation was set equal to 180° and you solved for the larger angle as well.

450. a. Let x = the width of the deck. Since the width of the pool only is 18 ft, the width of the pool and the deck together is $x + x + 18$, or $2x + 18$. Since the length of the pool only is 24 ft, the length of the pool and the deck together is $x + x + 24$, or $2x + 24$. The total area of the pool and the deck together is 832 square feet, which is 400 square feet for the deck added to 432 square feet for the pool. Area of a rectangle is *length* × *width*, so multiply the expressions together and set them equal to the total area of 832 square feet: $(2x + 18)(2x + 24) = 832$. Multiply the binomials using the distributive property: $4x^2 + 36x + 48x + 432 = 832$. Combine like terms: $4x^2 + 84x + 432 = 832$. Subtract 832 from both sides: $4x^2 + 84x + 432 - 832 = 832 - 832$; simplify: $4x^2 + 84x - 400 = 0$. Factor the trinomial completely: $2(2x^2 + 42x - 200) = 0$; $2(2x - 8)(x + 25) = 0$. Set each factor equal to zero and solve: $2 \neq 0$ or $2x - 8 = 0$ or $x + 25 = 0$; $x = 4$ or $x = {}^-25$. Reject the negative solution because a deck cannot have a negative width. The width of the deck is 4 feet.

451. b. The sum of the measures of the exterior angles of any polygon is 360°. Therefore, if the sum of four of the five angles equals 320°, to find the fifth simply subtract 320° from 360°, which equals 40°. If you chose **a**, you divided 325° by 4, assuming all four angles are equal in measure and assigned this value to the fifth angle, $\angle E$.

452. c. This problem requires two steps. First, determine the length of the base and height of the triangle. Second, determine the area of the triangle. To determine the base and height, we will use the equation $x + 4x = 95$. Simplifying, $5x = 95$. Divide both sides by 5: $x = 19$ cm. By substitution, the height is 19 cm and the base is 4(19) or 76 cm. The area of a triangle is found by using the formula $area = \frac{1}{2}\ base \times height$. Therefore, the area $= \frac{1}{2}(76)(19)$ or 722 cm². If you chose **a**, the area formula used was incorrect. Area $= \frac{1}{2}\ base \times height$, not $base \times height$. If you chose **b**, the original equation $x + 4x = 95$ was simplified incorrectly as $4x^2 = 95$.

453. c. First, convert all distances from meters to centimeters. Then, to solve for the height of the structure, solve the following proportion: $\frac{x}{10,000\ \text{cm}} = \frac{160\ \text{cm}}{800\ \text{cm}}$. Cross-multiply: $800x = 1,600,000$. Divide both sides by 800: $x = 2,000$. If you chose **b** or **d**, you made a decimal error.

454. c. In parallelogram $ABCD$, $\angle 2$ is equal in measurement to $\angle 5$. $\angle 2$ and $\angle 5$ are alternate interior angles, which are congruent. If $\angle 6$ is 120°, then $m\angle 6 + m\angle 5 + m\angle 4 = 180°$. Adjacent angles in a parallelogram are supplementary. Therefore, $40° + 120° + x = 180°$. Simplifying, $160° + x = 180°$. Subtract 160° from both sides: $x = 20°$. If you chose **a**, you assumed that $m\angle 4 + m\angle 5 = 90°$. If you chose **d**, you assumed that $m\angle 4$ is $\frac{1}{2}(m\angle 4 + m\angle 5)$.

455. c. To solve this problem, find the width of the frame first. Let $x =$ the width of the frame. Since the width of the picture only is 12 in, the width of the frame and the picture together is $x + x + 12$, or $2x + 12$. Since the length of the picture only is 14 in, the length of the frame and the picture together is $x + x + 14$, or $2x + 14$. The total area for the frame and the picture together is 288 square inches. Area of a rectangle is $length \times width$, so multiply the expressions together and set them equal to the total area of 288 square inches: $(2x + 12)(2x + 14) = 288$. Multiply the binomials using the distributive property: $4x^2 + 28x + 24x + 168 = 288$. Combine like terms: $4x^2 + 52x + 168 = 288$. Subtract 288 from both sides: $4x^2 + 52x + 168 - 288 = 288 - 288$; simplify: $4x^2 + 52x - 120 = 0$. Factor the trinomial completely: $4(x^2 + 13x - 30) = 0$; $4(x - 2)(x + 15) = 0$. Set each factor equal to zero and solve: $4 \neq 0$ or $x - 2 = 0$ or $x + 15 = 0$; $x = 2$ or $x = {}^-15$. Reject the negative solution because a frame cannot have a negative width. The width of the frame is 2 inches. Therefore, the larger dimension of the frame is $2(2) + 14 = 4 + 14 = 18$ inches.

456. d. The volume of a sphere is found by using the formula $\frac{4}{3}\pi r^3$. Since the volume is 288π cm^3 and we are asked to find the radius, we will set up the following equation: $\frac{4}{3}\pi r^3 = 288\pi$. To solve for r, multiply both sides by 3: $4\pi r^3 = 864\pi$. Divide both sides by π: $4r^3 = 864$. Divide both sides by 4: $r^3 = 216$. Take the cube root of both sides: $r = 6$ cm. If you chose **a**, the formula for volume of a sphere was incorrect; $\frac{1}{3}\pi r^3$ was used instead of $\frac{4}{3}\pi r^3$. If you chose **c**, you mistakenly took the square root of 216 instead of the cube root.

457. d. $\angle ABE$ and $\angle CBD$ are vertical angles that are equal in measurement. Solve the following equation for x: $4x + 5 = 7x - 13$. Subtract $4x$ from both sides; $5 = 3x - 13$. Add 13 to both sides: $18 = 3x$. Divide both sides by 3: $x = 6$. To solve for $m\angle ABE$ substitute $x = 6$ into the expression $4x + 5$ and simplify: $4(6) + 5 = 24 + 5$ or 29. $m\angle ABE$ equals 29°. If you chose **a**, you solved for $m\angle ABC$ or $m\angle EBD$. If you chose **b**, you assumed the angles were supplementary and set the sum of the two angles equal to 180°.

458. b. If two angles are complementary, the sum of the measurements of the angles is 90°. $m\angle 1$ is represented by x. $m\angle 2$ is represented by $4x$. Solve the following equation for x: $x + 4x = 90°$. Simplify: $5x = 90°$. Divide both sides by 5: $x = 18°$. The measure of the larger angle is $4x$ or $4(18)$, which equals 72°. If you chose **a**, the original equation was set equal to 180° rather than 90° and you solved for the smaller angle. If you chose **c**, the original equation was set equal to 180° rather than 90°, and you solved for the larger angle. If you chose **d**, you solved the original equation correctly; however, you solved for the smaller of the two angles.

459. d. To find how far the wheel will travel, find the circumference of the wheel and multiply it by 2. The formula for the circumference of a circle is πd. Since the diameter of the wheel is 25 inches, the circumference of the wheel is 25π. Multiply this by 2 to get $(2)(25\pi)$ or 50π. Finally, substitute 3.14 for π: $50(3.14) = 157$ inches, which is the distance the wheel travels in two turns. If you chose **a**, you used the formula for area of a circle rather than circumference. If you chose **b**, you calculated the distance traveled after one rotation, not two.

460. **a.** If two angles are supplementary, then the sum of the measurements of the angles is 180°. $m\angle 1$ is represented by x. $m\angle 2$ is represented by $2x + 30°$. Solve the following equation for x: $x + 2x + 30° = 180°$. Simplify: $3x + 30° = 180°$. Subtract 30° from both sides: $3x = 150°$. Divide both sides by 3: $x = 50°$. The measure of the larger angle is $2x + 30°$ or $2(50) + 30°$, which equals 130°. If you chose **b**, the equation was set equal to 90° rather than 180°, and you solved for the smaller angle. If you chose **c**, x was solved for correctly; however, this was the smaller of the two angles. If you chose **d**, the original equation was set equal to 90° rather than 180°, yet you continued to solve for the larger angle.

461. **d.** $m\angle AED$ and $m\angle BEC$ are vertical angles that are equal in measurement. Solve the following equation for x: $5x - 36 = 2x + 9$. Subtract $2x$ from both sides of the equation: $3x - 36 = 9$. Add 36 to both sides of the equation: $3x = 45$. Divide both sides by 3: $x = 15$. To solve for $m\angle AED$, substitute $x = 15$ into the expression $2x + 9$ and simplify: $2(15) + 9 = 39$. $m\angle AED$ equals 39°. If you chose **a**, you solved for the measure of, either $m\angle AEB$ or $m\angle DEC$. If you chose **b**, you assumed the angles were supplementary and set the sum of the angles equal to 180°. If you chose **c**, it was the same error as choice **b**.

462. **a.** The sum of the measures of the angles of a triangle is 180°. Using this fact, we can establish the following equation: $3x + 4x + 5x = 180°$. Simplify: $12x = 180°$. Divide both sides by 12: $x = 15$. The largest angle is represented by $5x$. Therefore, $5x$, or $5(15)$, equals 75°, which is the measure of the largest angle. If you chose **b**, the original equation was set equal to 90° rather than 180°. If you chose **c**, this was the smallest angle within the triangle. If you chose **d**, this is the angle whose measurement lies between the smallest and largest angles.

463. **c.** The widest piece of mail that is able to fit without bending will be equal to the length of the diagonal of the mailbox. The width, 4.5 in., will be one leg of the right triangle. The height, 5 in., will be the other leg of the right triangle. We will solve for the hypotenuse, which is the diagonal of the mailbox, using the Pythagorean theorem; $a^2 + b^2 = c^2$ or $4.5^2 + 5^2 = c^2$. Solve for c: $20.25 + 25 = c^2$; $45.25 = c^2$; $c = 6.7$ inches. If you chose **a**, you incorrectly assigned the legs the values of 4.5 and 10. If you chose **b**, you assigned the legs the values of 5 and 10.

464. d. To find the area of the cross section of pipe, we must find the area of the outer circle minus the area of the inner circle. To find the area of the outer circle, we will use the formula $Area = \pi r^2$. The outer circle has a diameter of $4(3 + \frac{1}{2} + \frac{1}{2})$ and a radius of 2; therefore, $Area = \pi 2^2$ or 4π. The inner circle has a radius of 1.5; therefore, $Area = \pi(1.5)^2$ or 2.25π. The difference, $4\pi - 2.25\pi$ or 1.75π in, is the area of the shaded region. If you chose **a**, you used the outer circle's radius of 3 and the inner circle's radius of $\frac{1}{2}$. If you chose **b** you used the outer circle's radius of $\frac{7}{2}$ and the inner circle's radius of 3. If you chose **c**, you used the outer radius of 4 and the inner radius of 3.

465. a. To find the volume of the pipe with a known cross section and length of 18 inches, simply multiply the area of the cross section by the length of the pipe. The area of the cross section obtained from the previous question was 1.75π in². The length is given as 18 inches. Therefore, the volume of the shaded region is 1.75 in² × 18 inches or 31.5π in³. If you chose **b**, you multiplied choice **c** from the previous question by 18. If you chose **c**, you multiplied choice **a** from the previous question by 18. If you chose **d**, you multiplied choice **b** from the previous question by 18.

466. c. Sketching an illustration is helpful for solving this problem. Observe in the diagram below that point A is the starting point and point B is the ending point. After sketching the four directions, we connect point A to point B. We can add to the illustration the total distance traveled north as well as the total distance traveled east. This forms a right triangle, given the distance of both legs, with the hypotenuse to be solved for. Using the Pythagorean theorem, $a^2 + b^2 = c^2$, or $8^2 + 15^2 = c^2$: $64 + 225 = c^2$; $289 = c^2$; $c = 17$ miles. If you chose **a**, you mistakenly traveled 4 miles due east instead of due west. If you chose **b**, you labeled the triangle incorrectly by assigning 15 to the hypotenuse rather than a leg.

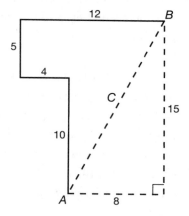

467. b. The area of the shaded region is the area of a rectangle measuring 22
units by 12 units, minus the area of a circle with a diameter of 12 units.
The area of the rectangle is (22)(12) = 264 square units. The area of a
circle with diameter 12, and a radius of 6, is $\pi(6)^2 = 36\pi$ square units. The
area of the shaded region is $264 - 36\pi$ square units. If you chose **a**, the
formula used for area of a circle was incorrect, $\frac{1}{2}\pi r^2$. If you chose **c**, the
formula used for area of a circle was incorrect, πd. If you chose **d**, this was
the reverse of choice **a**—area of the circle minus area of the rectangle.

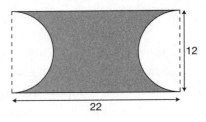

468. c. To find the area of the label, we will use the formula for the surface area
of a cylinder, *Surface Area* = πdh, which excludes the top and bottom of
the can. Substituting $d = 20$ and $h = 45$, the area of the label is $\pi(20)(45)$
= 900π cm². If you chose **a**, you used an incorrect formula for surface
area, *Surface Area* = πrh. If you chose **b**, you used an incorrect formula
for area, *Surface Area* = $\pi r^2 h$.

469. c. Let x = the width of the dictionary (in inches). Then, the height is
$(x + 2)$ inches. Since the perimeter is 48 inches and the dictionary is
rectangular, we use the fact that the perimeter of a rectangle is $2l + 2w$,
where l is the length and w is the width, to obtain the following equation: $2x + 2(x + 2) = 48$. To solve for x, simplify the left side: $2x + 2x +$
$4 = 48$, which is equivalent to $4x + 4 = 48$. Next, subtract 4 from both
sides: $4x = 44$. Finally, divide both sides by 4: $x = 11$. Therefore, the
dimensions of a typical page are 11 inches by 13 inches, and the area
of the page is 11 in. × 13 in. = 143 square inches.

470. d. Two parallel lines cut by a transversal form corresponding angles that
are congruent, or equal in measurement. $m\angle BAE$ is corresponding
to $m\angle CFE$. Therefore $m\angle CFE = 46°$. $m\angle CDF$ is corresponding to
$m\angle BEF$. Therefore, $m\angle BEF = 52°$. The sum of the measures of the
angles within a triangle is 180°. $m\angle CFE + m\angle BEF + m\angle FGE = 180°$.
Using substitution, $46° + 52° + m\angle FGE = 180$. Simplify: $98° + m\angle FGE$
$= 180°$. Subtract 98° from both sides; $m\angle FGE = 82°$. $m\angle FGE$ and

$m\angle CGE$ are supplementary angles. If two angles are supplementary, then the sum of their measurements equals 180°. Therefore, $m\angle FGE + \angle CGE$ = 180°. Using substitution, 82° + $m\angle CGE$ = 180°. Subtract 82° from both sides; $m\angle CGE$ = 98°. If you chose **a**, you solved for $m\angle CFE$. If you chose **b**, you solved for $m\angle BEF$. If you chose **c**, you solved for $m\angle FGE$.

471. b. To find the area of the shaded region, we must find the area of the circle minus the area of the rectangle. The formula for the area of a circle is πr^2. The radius is $\frac{1}{2}\overline{BC}$ or $\frac{1}{2}(10)$, which is 5 units. The area of the circle is $\pi(5)^2$ or 25π square units. The formula for the area of a rectangle is *length* × *width*. Using the fact that the rectangle can be divided into two triangles with a width of 6 and a hypotenuse of 10, and using the Pythagorean theorem, we will find the length: $a^2 + b^2 = c^2$; $a^2 + 6^2 = 10^2$; $a^2 + 36 = 100$; $a^2 = 64$; $a = 8$ units. The area of the rectangle is *length* × *width* or 6 × 8 = 48 square units. Finally, to answer the question, the area of the shaded region is the area of the circle – the area of the rectangle, or 25π – 48 square units. If you chose **a**, the error was in the use of the Pythagorean theorem, $6^2 + 10^2 = c^2$. If you chose **c**, the error was in finding the area of the rectangle. If you chose **d**, you used the wrong formula for area of a circle, πd^2.

472. b. The area of the shaded region is equal to the area of the square minus the area of the two semicircles. The area of the square is s^2 or 6^2, which equals 36 square units. The area of the two semicircles is equal to the area of one circle. *Area* = πr^2 or $\pi(3)^2$ or 9π square units. Therefore, the area of the shaded region is 36 – 9π square units. If you chose **a**, you calculated the area of the square incorrectly as 12. If you chose **c**, you used an incorrect formula for the area of two semicircles, $\frac{1}{2}\pi r^2$.

473. d. To solve for the length of the belt, begin with the distance from the center of each pulley, 3 ft, and multiply by 2: (3)(2) or 6 ft. Secondly, you need to know that the distance around two semicircles with the same radius is equivalent to the circumference of one circle. Therefore $C = \pi d$ or 12π inches. Since the units are in feet, and not inches, convert 12π inches to feet or 1π ft. Now add these two values together, $(6 + \pi)$ ft, to determine the length of the belt around the pulleys. If you chose **a** or **b**, you used an incorrect formula for circumference of a circle. Recall: *Circumference* = πd. If you chose **c**, you forgot to convert the unit from inches to feet.

474. d. To find the measure of an angle of any regular polygon, we use the formula $\frac{n-2}{n} \times 180°$, where n is the number of sides. Using 15 as the value for n, $\frac{15-2}{15} \times 180° = \frac{13}{15} \times 180°$ or 156°. If you chose **a**, you simply divided 360° (which is the sum of the exterior angles) by 15. If you chose **b**, you divided 180° by 15.

475. c. To find how many cubic yards of sand are in the pile, we must find the volume of the pile in cubic feet and convert the answer to cubic yards. The formula for volume of a cone is $V = \frac{1}{3}(height)(Area\ of\ the\ base)$. The area of the circular base is found by using the formula $Area = \pi r^2$. The area of the base of the sand pile is $\pi(16)^2$ or 803.84 ft². The height of the pile is 20 feet. The volume of the pile in cubic feet is $\frac{1}{3}(803.84)(20)$ or 5,358.93 ft³. To convert to cubic yards, divide 5,358.93 by 27 because 1 yard = 3 feet, and 1 yd³ means 1 yd × 1 yd × 1 yd which equals 3 ft × 3 ft × 3 ft or 27 ft³. The volume is 198.5 yd³. If you chose **a**, you did not convert to cubic yards. If you chose **b**, you converted incorrectly by dividing 5,358.93 by 9 rather than 27. If you chose **d**, the area of the base formula was incorrect. Area of a circle does not equal πd^2.

476. b. Observe below that the octagon can be subdivided into 8 congruent triangles. Since each triangle has a base of 4 and a height of 7, the area of each triangle can be found using the formula $Area = \frac{1}{2}base \times height$. To find the area of the octagon, we will find the area of one triangle and multiply it by 8. The area of one triangle is $\frac{1}{2}(4)(7)$ or 14 square units. Multiply this value by 8: (14)(8) = 112 square units. This is the area of the octagon. If you chose **a**, you used an incorrect formula for area of a triangle. $Area = base \times height$ was used rather than $area = \frac{1}{2}\ base \times height$. If you chose **d**, you mistakenly divided the octagon into 6 triangles instead of 8 triangles.

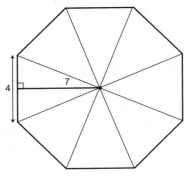

477. d. The sum of the measures of the angles of a quadrilateral is 360°. In the quadrilateral shown, three of the four angle measurements are known. They are 45° and two 90° angles. To find $m\angle A$, subtract these three angles from 360°: 360° − 90° − 90° − 45° = 135°. This is the measure of angle A. If you chose **a**, you assumed $\angle A$ and the 45° angle are complementary angles.

478. a. To find the total area of the shaded region, we must find the area of the rectangle minus the sum of the areas of all twelve circles. The area of the rectangle is *length × width*. Since the rectangle is 4 circles long and 3 circles wide, and each circle has a diameter of 10 cm (radius of 5 cm × 2), the rectangle is 40 cm long and 30 cm wide: (40)(30) = 1,200 cm². The area of one circle is πr^2 or $\pi(5)^2$ = 25π cm. Multiply this value times 12 because we are finding the area of 12 circles: (12)(25π) = 300π cm. The difference is 1,200 − 300π cm², which is the area of the shaded region. If you chose **b**, the area of the rectangle was incorrectly calculated as (20) (15). If you chose **c**, you reversed the area of the circles minus the area of the rectangle. If you chose **d**, you reversed choice **b** as the area of the circles minus the area of the rectangle.

479. c. Referring to the illustration below, $m\angle NEB$ = 23° and $m\angle DES$ = 48°. $m\angle NEB + m\angle BED + m\angle DES$ = 180°; using substitution, 23° + $m\angle BED$ + 48° = 180°. Simplify: 72° + $m\angle BED$ = 180°. Subtract 72° from both sides: $m\angle BED$ = 109°. If you chose **a**, you added 23 + 48 to total 71. If you chose **b**, you assumed $m\angle BED = m\angle NEB$. If you chose **d**, you assumed $m\angle BED = m\angle DES$.

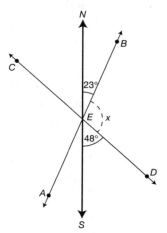

480. **a.** The measure of an angle of a regular polygon of n sides is $\frac{n-2}{n} \times 180°$. Since a hexagon has 6 sides, to find the measure of $\angle ABC$, substitute $n = 6$ and simplify. The measure of $\angle ABC$ is $\frac{6-2}{6} \times 180°$ or $120°$. If you chose **b**, you assumed a hexagon has 8 sides. If you chose **c**, you assumed a hexagon has 5 sides. If you chose **d**, you assumed a hexagon has 10 sides.

481. **d.** The volume of a box is found by multiplying *length* × *length* × *length* or $l \times l \times l = l^3$. If the length is doubled, the new volume is $(2l) \times (2l) \times (2l)$ or $8(l^3)$. When we compare the two expressions, we can see that the difference is a factor of 8. Therefore, the volume has been increased by a factor of 8.

482. **a.** The formula for finding the circumference of a circle is πd. If the radius is tripled, the diameter is also tripled. The new circumference is $\pi 3d$. Compare this expression to the original formula; with a factor of 3, the circumference is multiplied by 3.

483. **a.** The formula for the surface area of a sphere is $4\pi r^2$. If the diameter is doubled, this implies that the radius is also doubled. The formula then becomes $4\pi(2r)^2$. Simplifying this expression, $4\pi(4r^2)$ equals $16\pi r^2$. Compare $4\pi r^2$ to $16\pi r^2$; $16\pi r^2$ is 4 times greater than $4\pi r^2$. Therefore, the surface area is four times as large.

484. **b.** If the diameter of a sphere is tripled, then the radius is also tripled. The formula for the volume of a sphere is $\frac{4}{3}\pi r^3$. If the radius is tripled, *volume* $= \frac{4}{3}\pi(3r)^3$ which equals $\frac{4}{3}\pi(27r^3)$ or $\frac{4}{3}(27)\pi r^3$. Compare this equation for volume with the original formula; with a factor of 27, the volume is now 27 times as large.

485. **b.** The formula for the volume of a cone is $\frac{1}{3}\pi r^2 h$. If the radius is doubled, then *Volume* $= \frac{1}{3}\pi(2r)^2 h$ or $\frac{1}{3}\pi 4r^2 h$. Compare this expression to the original formula; with a factor of 4, the volume is now 4 times as large.

486. **a.** The formula for the volume of a cone is $\frac{1}{3}\pi r^2 h$. If the radius is halved, the new formula is $\frac{1}{3}\pi(\frac{1}{2}r)^2 h$ or $\frac{1}{3}\pi(\frac{1}{4})r^2 h$. Compare this expression to the original formula; with a factor of $\frac{1}{4}$, the volume is now $\frac{1}{4}$ as large.

487. b. The volume of a right cylinder is $\pi r^2 h$. If the radius is doubled and the height is halved, the new volume is $\pi(2r)^2(\frac{1}{2}h)$ or $\pi 4r^2(\frac{1}{2}h)$ or $2\pi r^2 h$. Compare this expression to the original formula; with a factor of 2, the volume is now 2 times as large.

488. a. The formula for the volume of a right cylinder is *Volume* = $\pi r^2 h$. If the radius is doubled and the height is tripled, the new volume is $\pi(2r)^2(3h)$. Simplified, this becomes $\pi 4r^2 3h$ or $\pi 12r^2 h$. Compare this expression to the original formula; with a factor or 12, the volume is now 12 times as large.

489. c. The measure of an angle of a regular polygon of n sides is $(n - n\ 2) \times 180°$. Since each angle measures 150°, we will solve for n, the number of sides. Using the formula $150° = (n - n\ 2) \times 180°$, solve for n. Multiply both sides by n,: $150°n = (n - 2)180°$. Distribute 180°: $150°n = 180°n - 360°$. Subtract $180n$ from both sides: $^-30°n = ^-360°$. Divide both sides by $^-30$: $n = 12$. The polygon has 12 sides. Alternately, if each angle is 150°, each exterior angle is $180° - 150° = 30°$. The measure of an exterior angle of a rectangular polygon of n sides is $\frac{360°}{n}$. We now solve for n by solving $\frac{360°}{n} = 30°$. Multiplying both sides by n, $360° = 30°n$. Divide both sides by 30°: $n = 12$.

490. d. This problem requires two steps. First, find the diagonal of the base of the box. Second, using this value, find the length of the diagonal $\overline{AB}$. To find the diagonal of the base, use 30 cm as a leg of a right triangle and 8 cm as the second leg, and solve for the hypotenuse. Using the Pythagorean theorem: $30^2 + 8^2 = c^2$; $900 + 64 = c^2$; $964 = c^2$; $c = 31.05$ cm. Now consider this newly obtained value as a leg of a right triangle and 12 cm as the second leg, and solve for the hypotenuse, $\overline{AB}$: $31.05^2 + 12^2 = \overline{AB}^2$: $964 + 144 = \overline{AB}^2$: $1{,}108 = \overline{AB}^2$. $\overline{AB} = 33.3$ cm. If you chose **a**, you used 30 and 12 as the measurements of the legs. If you chose **b**, you solved the first triangle correctly; however, you used 8 as the measure of one leg of the second triangle, which is incorrect.

491. c. To find the area of the shaded region, we must find $\frac{1}{2}$ the area of the circle with diameter AC, minus $\frac{1}{2}$ the area of the circle with diameter BC, plus $\frac{1}{2}$ the area of the circle with diameter AB. To find $\frac{1}{2}$ the area of the circle with diameter AC, we use the formula $Area = \frac{1}{2}\pi r^2$. Since the diameter is 6, the radius is 3; therefore, $Area = \frac{1}{2}\pi(3)^2$ or 4.5π square units. To find $\frac{1}{2}$ the area of the circle with diameter BC, we again use the formula $Area = \frac{1}{2}\pi r^2$. Since the diameter is 4, the radius is 2; therefore $Area = \frac{1}{2}\pi(2)^2$ or 2π square units. To find $\frac{1}{2}$ the area of the circle with diameter AB we use the formula $Area = \frac{1}{2}\pi r^2$. Since the diameter is 2, the radius is 1; therefore the area is $\frac{1}{2}\pi$ square units. Finally, $4.5\pi - 2\pi + 0.5\pi = 3\pi$ square units, which is the area of the shaded region. If you chose **a** or **b**, in the calculations you mistakenly used πd^2 as the area formula rather than πr^2.

492. a. This problem has three parts. First, we must find the diameter of the tower. Secondly, we will increase the diameter by 16 meters for the purpose of the fence. Finally, we will find the circumference using this new diameter. This will be the length of the fence. The formula for the circumference of a circle is πd. This formula, along with the fact that the tower has a circumference of 40 meters, gives us the following formula: $40 = \pi d$. To solve for d, the diameter, divide both sides by π. $d = \frac{40}{\pi}$, which is the diameter of the tower. Now increase the diameter by 16 meters; $\frac{40}{\pi} + 16$ is the diameter of the fenced-in section. Finally, use this value for d in the equation *circumference* = πd or $\pi(\frac{40}{\pi} + 16)$ meters. Simplify by distributing π through the expression: $(40 + 16\pi)$ meters. This is the length of the security fence. If you chose **b**, you added 8 to the circumference of the tower rather than 16. If you chose **c**, you merely added 8 to the circumference of the tower.

493. a. Using the illustration below, $m\angle 2 = m\angle a$. because they are vertical angles. $\angle 1$ and $\angle a$ are supplementary, since $m\angle c + m\angle d + m\angle 1 + m\angle a = 360°$ (the total number of degrees in a quadrilateral): $90° + 90° + m\angle 1 + m\angle a = 360°$. Simplifying further, $180° + m\angle 1 + m\angle a = 360°$. Subtract $180°$ from both sides: $m\angle 1 + m\angle a = 180°$. Since $m\angle a = \angle 2$, using substitution, $m\angle 1 + m\angle 2 = 180°$. Using similar logic, $m\angle 4 = m\angle b$ because they are vertical angles. $m\angle 3$ and $m\angle b$ are supplementary because, using the same process described above, $m\angle b + m\angle 3 = 180°$. Since $m\angle b = m\angle 4$, using substitution, $m\angle 3 + m\angle 4 = 180°$. Finally, adding $m\angle 1 + m\angle 2 = 180°$ to $m\angle 3 + m\angle 4 = 180°$, we can conclude $m\angle 1 + m\angle 2 + m\angle 3 + m\angle 4 = 360°$.

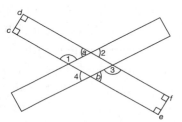

494. a. To find the volume of the hollow solid, we must find the volume of the original cone minus the volume of the smaller cone sliced from the original cone minus the volume of the cylindrical hole. The volume of the original cone is found by using the formula $V = \frac{1}{3}\pi r^2 h$. Using the values $r = 9$ and $h = 40$, substitute and simplify to find $Volume = \frac{1}{3}\pi(9)^2(40)$ or $1{,}080\pi$ cm^3. The volume of the smaller cone is found by using the formula $V = \frac{1}{3}\pi r^2 h$. Using the values $r = 3$ and $h = 19$, substitute and simplify to find $Volume = \frac{1}{3}\pi(3)^2(19)$ or 57π cm^3. The $Volume$ of the cylinder is found by using the formula $V = \pi r^2 h$. Using the values $r = 3$ and $h = 21$, substitute and simplify to find $Volume = \pi(3)^2(21)$ or 189π cm^3. Finally, calculate the volume of the hollow solid: $1{,}080\pi - 57\pi - 189\pi = 834\pi$ cm^3. If you chose **b**, you used an incorrect formula for the volume of a cone, $V = \pi r^2 h$. If you chose **c**, you subtracted the volume of the large cone minus the volume of the cylinder. If you chose **d**, you added the volumes of all three sections.

495. c. To find the volume of the object, we must find the volume of the water that is displaced after the object is inserted. Since the container is 5 cm wide and 15 cm long, and the water rises 2.3 cm after the object is inserted, the volume of the displaced water can be found by multiplying length by width by depth: $(5)(15)(2.3) = 172.5$ cm^3.

496. a. To find how many cubic yards of concrete are needed to construct the wall, we must determine the volume of the wall. The volume of the wall is calculated by finding the surface area of the end and multiplying it by the length of the wall, 120 ft. The surface area of the end of the wall is found by dividing it into three regions, calculating each region's area, and adding them together. The regions are labeled A, B, and C. To find the area of region A, multiply the length (3) times the height (10) for an area of 30 ft^2. To find the area of region B, multiply the length (5) times the height (3) for an area of 15 ft^2. To find the area of region C, multiply $\frac{1}{2}$ times the base (2) times the height (4) for an area of 4 ft^2. The surface area of the end is 30 ft^2 + 15 ft^2 + 4 ft^2 = 49 ft^2. Multiply 49 ft^2 by the length of the wall 120 ft; 5,880 ft^3 is the volume of the wall. The question, however, asks for the answer in cubic yards. To convert cubic feet to cubic yards, divide 5,880 ft^3 by 27 ft^3, the number of cubic feet in one cubic yard, which equals 217.8 yd^3. If you chose **b**, you did not convert to yd^3. If you chose **c**, the conversion to cubic yards was incorrect. You divided 5,880 by 9 rather than 27. If you chose **d**, you found the area of the end of the wall and not the volume of the wall.

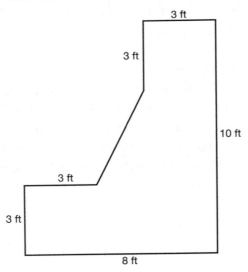

497. b. To find the volume of the sphere we must find the volume of the outer sphere minus the volume of the inner sphere. The formula for the volume of a sphere is $\frac{4}{3}\pi r^3$. The volume of the outer sphere is $\frac{4}{3}\pi(120)^3$. Here the radius is 10 feet (half the diameter) multiplied by 12 (converted to inches), or 120 inches. The volume equals 7,234,560 in³. The volume of the inner sphere is $\frac{4}{3}\pi(119)^3$ or 7,055,199. (This is rounded to the nearest cubic foot.) The difference of the volumes is 7,234,560 – 7,055,199 or 179,361 in³. This answer is in cubic inches, and the question asks for cubic feet. Since one cubic foot equals 1,728 cubic inches, we simply divide 179,361 by 1,728, which equals 104 ft, rounded to the nearest cubic feet. As an alternative to changing units to inches only to have to change them back into feet again, keep units in feet. The radius of the outer sphere is 10 feet, and the radius of the inner sphere is one inch less than 10 feet, which is $9\frac{11}{12}$ feet, or 9.917 feet. Use the formula for volume of a sphere, $\frac{4\pi r^3}{3}$, and find the difference in the volumes. If you chose **a**, you used an incorrect formula for the volume of a sphere, $V = \pi r^3$. If you chose **c**, you also used an incorrect formula for the volume of a sphere, $V = \frac{1}{3}\pi r^3$. If you chose **d**, you found the correct answer in cubic inches; however, your conversion to cubic feet was incorrect.

498. c. To solve this problem, we must find the volume of the sharpened tip and add this to the volume of the remaining lead that has a cylindrical shape. To find the volume of the sharpened point, we will use the formula for finding the volume of a cone, $\frac{1}{3}\pi r^2 h$. Using the values $r = 0.0625$ (half the diameter) and $h = 0.25$, $Volume = \frac{1}{3}\pi(0.0625)^2(0.25) = 0.002$ in³. To find the volume of the remaining lead, we will use the formula for finding the volume of a cylinder, $\pi r^2 h$. Using the values $r = 0.0625$ and $h = 5$, $Volume = \pi(0.0625)^2(5) = 0.0613$. Therefore, the volume of the lead is $0.001 + 0.0613 = 0.0623$ in³, which is 0.062 when rounded to the nearest thousandth. If you chose **a**, this is the volume of the lead without the sharpened tip. If you chose **b**, you subtracted the volumes calculated.

499. b. To find the volume of the hollow solid, we must find the volume of the cube minus the volume of the cylinder. The volume of the cube is found by multiplying *length* × *width* × *height* = (5)(5)(5) = 125 in³. The volume of the cylinder is found by using the formula $\pi r^2 h$. The radius of the cylinder is 2, and the height is 5. Therefore, the volume is $\pi(2)^2(5)$ or 20π in. The volume of the hollow solid is $125 - 20\pi$ in. If you chose **a**, you made an error in the formula of a cylinder, using $\pi d^2 h$ rather than $\pi r^2 h$. If you chose **c**, you reversed the volumes in choice **a**. This is the volume of the cylinder minus the volume of the cube. If you chose **d**, you found the reverse of choice **b**.

500. b. Refer to the diagram to find the area of the shaded region. One method is to enclose the figure into a rectangle, and subtract the area of the unwanted regions from the area of the rectangle. The unwanted regions have been labeled *A* through *F*. The area of region *A* is (15)(4) = 60 units². The area of region *B* is (5)(10) = 50 units². The area of region *C* is (20)(5) = 100 units². The area of region *D* is (17)(3) = 51 units². The area of region *E* is (20)(5) = 100 units². The area of region *F* is (10)(5) = 50 units². The area of the rectangle is (23)(43) = 989 units². Therefore, the area of the shaded region is 989 – 60 – 50 – 100 – 51 – 100 – 50 = 578 units². If you chose **a**, **c** or **d**, you omitted one or more of the regions *A* through *F*.

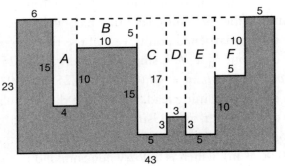

501. d. The shape formed by the paths of the two arrows and the radius of the bull's eye is a right triangle. The radius of the bull's eye is one leg, and the distance the second arrow traveled is the second leg. The distance the first arrow traveled is the hypotenuse. To find the distance the first arrow traveled, use the Pythagorean theorem, where 2 meters (half the diameter of the target) and 20 meters are the lengths of the legs and the length of the hypotenuse is missing. Therefore, $a^2 + b^2 = c^2$ and $a = 2$ and $b = 20$, so $2^2 + 20^2 = c^2$. Simplify: $4 + 400 = c^2$. Simplify: $404 = c^2$. Find the square root of both sides: $20.1 = c$. So the first arrow traveled about 20.1 meters. If you chose **c**, you added the two lengths together without squaring. If you chose **b**, you added Kim's distance from the target to the diameter of the target. If you chose **a**, you let 20 meters be the hypotenuse of the right triangle instead of a leg, and you used the radius of the target.

Additional Online Practice

Whether you need help building basic skills or preparing for an exam, visit the LearningExpress Practice Center! On this site, you can access additional practice materials. Using the code below, you'll be able to log in and access additional online math word problems practice. This one-time use online practice will also provide you with:

- **Immediate scoring**
- **Detailed answer explanations**
- **Personalized recommendations for further practice and study**

Log in to the LearningExpress Practice Center by using this URL: **www.learnatest.com/practice**

This is your Access Code: **9049**

Follow the steps online to redeem your access code. After you've used your access code to register with the site, you will be prompted to create a username and password. For easy reference, record them here:

Username: _____ Password: _____

With your username and password, you can log in and access the additional practice material. If you have any questions or problems, please contact LearningExpress customer service at 1-800-295-9556 ext.2, or e-mail us at **customerservice@learningexpressllc.com**.